Lab Manual

TO ACCOMPANY EXPLORING AGRISCIENCE

Fifth Edition

Lab Manual TO ACCOMPANY EXPLORING AGRISCIENCE

Fifth Edition

Ray V. Herren
Alicia Tomlinson

CENGAGE Learning

Australia • Brazil • Mexico • Singapore • United Kingdom • United States

CENGAGE Learning

Lab Manual to Accompany Exploring Agriscience, 5E
Ray V. Herren

SVP, GM Skills & Global Product Management: Jonathan Lau

Product Director: Matt Seeley

Product Manager: Nicole Robinson

Senior Director, Development: Marah Bellegarde

Senior Product Development Manager: Larry Main

Senior Content Developer: Jennifer Starr

Product Assistant: Deb Handy

Vice President, Marketing Services: Jennifer Ann Baker

Marketing Manager: Kelsey Hagan

Senior Production Director: Wendy Troeger

Production Director: Andrew Crouth

Senior Content Project Manager: Betsy Hough

Senior Art Director: Benj Gleeksman

Cover images: Lab: iStockPhoto.com/mediaphotos; Cows: iStockPhoto.com/emholk

Chicks: iStockPhoto.com/danchooalex; Oranges: Marcie Gonzalez/Collection/Getty Images

Library of Congress Control Number: 2016935735

ISBN: 978-1-305-94983-6

Cengage Learning
20 Channel Center Street
Boston, MA 02210
USA

Printed at CLDPC, USA, 08-24

Contents

Note to Students

The exercises included in this lab manual are designed to allow you to experience and practice procedures used by agricultural and biological scientists. One of the most important of these is the scientific method, which scientists use as a model for problem-solving to develop solutions to real-world problems. Many of the lab exercises ask you to use the scientific method including recording data on a daily or repeated basis, analyzing the results of your experiments, and making recommendations about practical solutions to the problem. Many of the experiments could be expanded for science fair projects, 4-H, FFA, or agricultural competitions. It is our hope that you will enjoy applying the knowledge gained in the agricultural science classroom to experiments and problem solving.

A Word about Lab Safety

Safety in the lab is of the upmost importance. Before participating in experiments in the classroom, it is important to review the following important safety points:

- Start at the beginning. Understand the safety concerns of the lab environment before it becomes a hazardous situation. The first exercise conducted in the classroom should be *An Introduction to Lab Safety* from the accompanying *Exploring Agriscience Lab Manual.*
- Chemicals and equipment are used in all science labs. With proper handling and technique, the equipment, organisms, and chemicals used in these lab exercises are safe.
- It is important that you follow all directions carefully. Pay particular attention to CAUTION notes where they appear. The CAUTION notes contain special precautions for the handling of chemicals and living organisms.
- Do not eat or drink in the lab classroom and always wash your hands at the beginning and end of each class period.
- Wear safety goggles and use gloves when appropriate.
- Those individuals with hair below the ears should tie hair away from the face before working with any open flame (like an alcohol lamp or bunsen burner).
- Long pants that completely cover the legs and leather shoes that enclose the feet should be worn while completing the lab activities (no sandals).
- Some lab exercises require field trips for the collection of samples. Dress suitably for any fieldwork and follow all directions outlined by your teacher.

REMEMBER THAT SCIENCE IS FUN, BUT FOR EVERYONE TO BE SAFE, ALL RULES OF CONDUCT IN THE CLASSROOM MUST BE FOLLOWED!

Note to Teachers

Looking for the solutions and answers to the problems and questions found in this Lab Manual? Visit CengageBrain.com to access the secure Companion site to accompany *Science of Agriculture: A Biological Approach, 5E* and find this and other helpful teaching tools. New to CengageBrain.com? Go to http://login.cengage.com and follow the prompts for obtaining access this secure site.

About the Contributing Author

ALICIA TOMLINSON
A University of Georgia graduate in Animal Science, Alicia taught high school agriculture in Georgia for 7 years before starting her company,

One Less Thing in 2007. The mission of *One Less Thing* is to make teaching Ag easier by creating curriculum and teaching supplements that are simple and relevant for agriculture educators, including lesson plans, classroom activities, and Career Development Event study materials for middle and high school students. An Ag teacher at heart, Alicia continually strives to "make a difference" in the lives of not only the 4-H agents, teachers and parents that *One Less Thing* serve, but ultimately, in the lives of agriculture students across the country.

INTRODUCTION

Before You Begin: An Introduction to Lab Safety

Student Objectives

After completing this lab activity, you should be able to:

- Describe various pieces of personal safety equipment and how they are properly used in a lab exercise.
- Determine the importance of lab safety and how it relates to the successful completion of a lab exercise.
- Accurately demonstrate proper safety procedures.
- Practice safe operation of lab equipment during the setup, the lab exercise, and clean up.

Safety in and out of the classroom is an important part of every lab exercise. **Safety** is the practice of avoiding dangerous situations that could cause injury or loss in the process of completing a lab exercise. Your personal safety, as well as the safety of your classmates and instructor, is always of the utmost importance in the completion of the lab exercises. Each student must take time to read the safety procedures for each lab and become familiar with all classroom safety procedures presented by your instructor.

Lab Safety

There are several keys to lab safety that students must follow that will allow the class to complete a lab successfully. First and foremost, each student must develop safe work habits during each lab. Lab instruments, such as scalpels, watch glasses, and forceps, all have the ability to cause injury if they are not used for their intended purpose during a lab exercise. Lab equipment like microscopes and balances can also be damaged and injure students if they are not handled properly before, during, and after completion of the lab. By caring for the tools and instruments supplied during the lab exercises, students can develop good work habits that will lead to safe and effective labs for the entire class.

Personal safety during each class period is also a key component of lab safety. There are several different types of personal safety equipment that may be supplied to you during a lab. Particle masks and respirators are two items that are used to ensure each student breathes clean air during an activity that has harmful vapors (e.g., chlorine gas) or fine particles floating in the air (e.g., sawdust). Safety glasses, goggles, or face shields all provide different levels of eye and face protection from chemical splash or flying particles from equipment use. Earmuffs and earplugs provide ear protection from the engine noise of power equipment or heavy machinery used in an activity. Insulated gloves and clothing provide protection from heat during labs involving gas burners and hot plates. Beyond these personal safety items, students should avoid wearing loose clothing and jewelry that could become tangled during the lab. Students should also wear full-length pants and shoes that cover the entire foot to provide an additional level of personal safety.

CAUTION:
The CAUTION notations appear in several different exercises in this lab manual. They highlight proper use of sharp instruments; proper clothing for the exercise whether inside or outside of the classroom; and proper hand-washing and area clean-up before, during, and after the exercise. These notations should always be considered in preparation for the exercise: ignoring them can lead to injury or sickness.

Activity: Practicing Lab Safety

Materials Needed

- Lab equipment set up at various stations
- Cleaning materials (lens paper, cloths, cleaning solution)
- Ultraviolet light
- Glo-Germ powder

Procedure

1. Your teacher will have various pieces of lab equipment set out at each lab station in the classroom. You will be instructed to break up into your lab groups and go to your assigned station.
2. At your station, inspect all the equipment and tools as you would normally before beginning the lab. After inspecting each piece, thoroughly clean each item and return each to its appropriate storage area as directed by your instructor.
3. Once each group is finished cleaning all the lab items, your teacher will produce an ultraviolet light and explain the purpose of this lab. Your teacher has treated each lab item with a chemical called Glo-Germ™ powder that contains tiny pieces of plastic that simulate germs. These tiny, inert pieces of plastic glow under ultraviolet light.

4. Your teacher will walk around the class and pass the light over each student's hands. Any traces of the powder will glow under the light. After the teacher has inspected each student, he or she will then inspect all the lab equipment and lab station surfaces for traces of the powder.

Questions for Thought

1. Is it possible to remove every trace of the germ powder from every surface used in the lab exercise?

2. What safety equipment is useful in preventing **germ transmission** in a lab exercise?

3. Was there a particular piece of lab equipment or tool that was consistently free of germ powder among all the lab groups? Were there any items that none of the groups were able to completely clean? Describe.

4. What if this was actual bacteria, such as the Salmonella in exercise 16, what are the consequences of not washing your hands thoroughly and cleaning all the lab equipment?

5. How well would you perform if this germ powder was used while you prepared a meal for yourself at home?

Glossary

germ transmission: the process of transferring a microscopic organism from an infected substance to other people, equipment, or work surface

safety: the ability to avert or not cause injury, danger, or loss

EXERCISE 1

Units of Measurement: The Metric System

Student Objectives

After completing this lab activity, you should be able to:

- Make measurements using a metric ruler.
- Convert the metric units of millimeters (mm), centimeters (cm), and meters (m) to the U.S. units of inches, feet, and yards.
- Use a metric balance to weigh some common objects in the metric units of grams (g) and kilograms (kg).
- Convert the metric weights of some common objects to the U.S. units of ounces (oz) and pounds (lb).
- Use measuring cups with the metric units of fluid volume milliliters (ml) and liters (l) to determine the volume of some liquids.
- Convert the metric fluid volumes of some liquids to the U.S. units of pints, quarts, and gallons.
- Measure temperature in degrees Celsius (°C) and convert temperature to degrees Fahrenheit (°F).

Materials Needed:

- Calculator
- A ruler with both inches and centimeters marked
- A scale (triple-beam or digital)
- Cup of dried bean seeds
- Five different objects to weigh
- A thermometer with both Celsius (°C) and Fahrenheit (°F) scales

Recommended Resources:

Chapter 1 in *Exploring Agriscience, 5th Edition.*

Introduction

Everyone uses measurements—the weight of an animal, volumes of foods used in recipes, or the distance to the nearest house. There are two standardized methods of measuring: the U.S. Customary System and the metric system. In the United States, we generally measure distance based on the average length of the human **foot**, 12 inches. Measurements of weight and volume are familiar to you. When you step on a scale at the doctor's office, your weight is measured in **pounds** (lb), the standard U.S. unit of weight or mass. If you look at your box of breakfast cereal, the weight of the cereal is measured in **ounces** (oz) and in **grams**. In the U.S. system of measurement, 16 measured ounces is the same as one pound. If you help your parents make cookies, you might add one-half pound of chocolate chips (which is 8 ounces) to your cookie dough.

Other countries use an alternative way of measurement called the metric system. The **metric system** is based upon basic units which are multiplied or divided by 10. By using a system built on multiples of 10, it is very easy to change from one unit of measurement to another. Scientists and engineers use the metric system because it is the most common method of measurement throughout the world.

The U.S. Customary System is based on the British Imperial System of measurement. We measure length in **inches**, **feet**, yards, and miles. We measure weight in ounces and pounds, and measure liquid volume with pints, quarts, and gallons. Air, water, and soil temperature are all currently measured on the **Fahrenheit** temperature scale (°F), although many weather reports also provide air temperature in degrees **Celsius** (°C). In the United States, most producers and agricultural suppliers still use the U.S. Customary System of measurement. For example, diesel fuel is priced per gallon. Wheat, corn, and soybean yields are measured in bushels (units of dry volume), and livestock weight is measured in pounds. Products imported from Germany, France, Sweden, and Japan are built using metric nuts, bolts, and tools. If you buy a tractor built overseas and want to make repairs, you will have to use a metric set of wrenches. The United States is the only country that does not use the metric system exclusively. Most of us are comfortable with U.S. units of measurement, but scientists need a more standardized measurement system. Scientists need to be able to communicate with other scientists throughout the world. For this reason, scientific measurements are made using the metric system.

You need to be familiar with the metric system. One day you may need to repair a Japanese tractor using metric tools, compare crop yields in the United States with crop yields in a European country, or purchase animal medicines made overseas. All these jobs require a good working knowledge of the metric system. Let's begin our exploration by learning about the basic units of measurement. Only a single metric conversion for each unit of length, weight (mass), and volume is needed to be able to convert between the U.S. and metric systems.

Units of Length

Length is the distance between two points. The standard unit of length in the U.S. system is the inch. Its counterpart in the metric system is a **centimeter.** One inch is equal to 2.54 centimeters (abbreviated cm) and three feet (one yard) is equivalent to 0.9144 **meter** (abbreviated m). In math class you learned that 12 inches is equal to 1 foot, 3 feet is the same as 1 yard, and 1 mile is 5,280 feet. The conversion of inches to feet and feet to miles requires the use of odd conversion units. Metric conversions use multiples of 10 (see Table 1-1). You need to learn the metric prefixes, which are based on Greek and Latin terms for the units such as *centi-* for one one-hundredth, *milli-* for one one-thousandth, and so on.

How do we go about converting from U.S. units to metric units of length? Suppose you want to know how far it is from your house to school in **kilometers.** You know that you live 10 miles from school and that 2.54 centimeters is equal to 1 inch. You also know that there are 100 centimeters in a **meter** and 1,000 meters in a kilometer (see Table 1-1). You remember from math class that there are 12 inches in 1 foot and 5,280 feet in 1 mile. Your problem is solved as follows:

$$10 \text{ miles} \times \frac{(5.280 \text{ ft})}{(1 \text{ mile})} \times \frac{(12 \text{ in.})}{(1 \text{ ft})} \times \frac{(2.54 \text{ cm})}{(1 \text{ in.})} \times \frac{(1 \text{ m})}{(100 \text{ cm})} \times \frac{(1 \text{ km})}{(1{,}000 \text{ m})} = \text{distance (km)}$$

If we cancel all units that appear in both the numerator (top) and the denominator (bottom) of the equation, we are left with the unit of interest, kilometers. This is one way to check to be certain that you have set up your equation correctly. We started with miles (our known distance) and have ended with kilometers.

$$10 \ \cancel{\text{miles}} \times \frac{(5.280 \ \cancel{\text{ft}})}{(1 \ \cancel{\text{mile}})} \times \frac{(12 \ \cancel{\text{in.}})}{(1 \ \cancel{\text{ft}})} \times \frac{(2.54 \ \cancel{\text{cm}})}{(1 \ \cancel{\text{in}})} \times \frac{(1 \ \cancel{\text{m}})}{(100 \ \cancel{\text{cm}})} \times \frac{(1 \text{ km})}{(1{,}000 \ \cancel{\text{m}})} = \text{distance (km)}$$

Prefix	Portion of Unit (meter, liter, gram)
Kilo-	1,000 or 10^3
Centi-	1/100 or 10^{-2} (0.01)
Milli-	1/1000 or 10^{-3} (0.001)
Micro-	1/1,000,000 or 10^{-6} (0.000001)
Nano-	1/1,000,000,000 or 10^{-9} (0.000000001)

Table 1-1. Some prefixes used in metric measurements.

Next, we multiply all the numbers in the numerator together, and multiply all the numbers in the denominator together.

$$\frac{10 \times 5{,}280 \times 12 \times 2.54 \times 1 \times 1}{1 \times 1 \times 1 \times 100 \times 1{,}000} = \frac{1{,}609{,}344}{100{,}000} \text{distance (km)}$$

The next step is to divide the top number (the numerator) by the bottom number (the denominator).

$$\frac{1{,}609{,}344}{100{,}000} = 16.09344 \text{ km}$$

The answer would be correctly expressed as 16.09 km. The answer is written with two numbers to the right of the decimal point. Since 0.00344 is less than 0.005, we round down to zero.

Using the ruler provided by your teacher, measure the following objects using the U.S. unit of length, the inch.

Width of your agriscience textbook = ____________ inches
Length of your pencil or pen = ____________ inches
Length of your shoe = ____________ inches
Width of the palm of your hand = ____________ inches

Using the conversions 1 in. = 2.54 cm and 100 cm = 1 m, convert these values from inches to centimeters.

Width of your agriscience textbook = ____________ centimeters
Length of your pencil or pen = ____________ centimeters
Length of your foot = ____________ centimeters
Width of the palm of your hand = ____________ centimeters

Using the metric side of your ruler, measure three objects provided by your teacher in millimeters (abbreviated mm). Fill in the values in the first column in Table 1-2. Convert the length of these objects to feet using the following conversions: 10 mm/cm, 2.54 cm/in., and 12 in./ft. Write your answers in the second column of Table 1-2.

Name of Object	Measurement (millimeters)	Measurement (feet)
1.		
2.		
3.		
4.		

Table 1-2. Measurements of four objects using the metric unit of length, millimeter, and the corresponding values in feet (English unit).

Units of Weight (Mass)

Mass is defined as the amount of substance in an object. We estimate mass by weight. You have had experience with the familiar U.S. units of mass or weight, the ounce (oz) and the pound (lb). One pound is equal to 16 ounces. Less familiar are the metric units of weight, the **gram** (abbreviated g) and the **kilogram** (abbreviated kg). If you look at Table 1-1, you will see that 1,000 grams equals 1 kilogram. Scientists who study physics, chemistry, and biology use grams and kilograms as units of weight. European, Latin American, and Asian countries also use the metric units for weight, as does Canada. Mass is typically measured using some type of scale. The scale you will be using to make measurements is similar to the scale used in many research labs and field stations where some degree of precision is needed, but not for very small quantities. The scale probably looks similar to the **triple-beam balance**, as shown in Figure 1-1.

To weigh accurately, it is important to be certain that the scale is "zeroed" before you take the first value. To zero a triple-beam balance, there is an adjustment or zero knob under the pan; turn the knob until the arrow on the right end of the arm is pointing to zero. On a digital scale, there is often a button marked "tare" which zeros the scale when pushed. It is very important that the scale is zeroed so that you do not over- or under-weigh the sample you are measuring. Your teacher will give you a measuring cup filled with bean seeds. You need to figure out how much the bean seeds weigh in grams. When you place the measuring cup full of seeds on the balance, you get the weight of the measuring cup and the seeds together.

How would you find the weight of the bean seeds if you cannot pour the seeds out on the balance pan? You would approximate the weight of the seeds by measuring the weight of the measuring cup first and then finding the weight of the measuring cup filled with seeds. You would then subtract the weight of the measuring cup from the weight of the measuring cup plus the seeds. The amount of weight left is the weight of the bean seeds alone. Scientists and engineers estimate

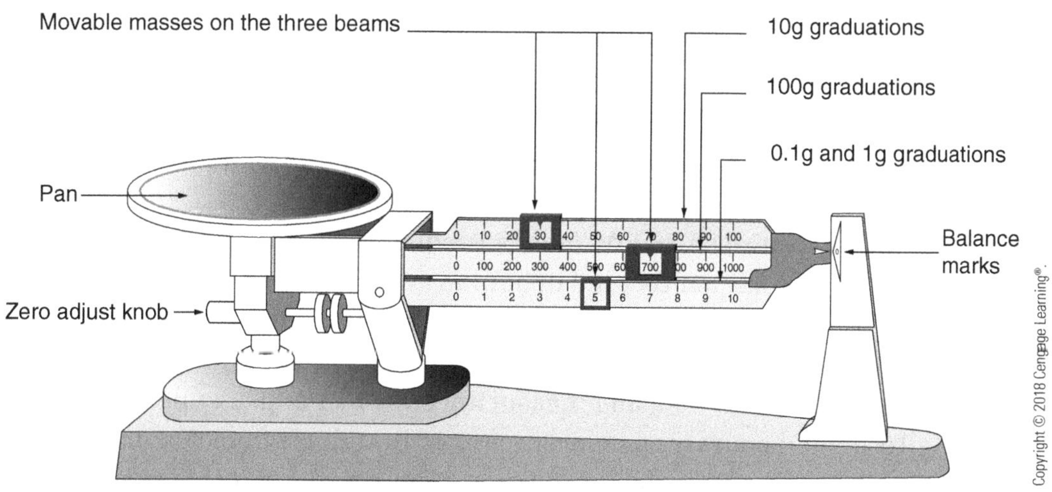

Figure 1-1. A triple beam balance.

Name of Object	Weight/Mass (in grams)	Weight/Mass (in kilograms)	Weight/Mass (in pounds)
1.			
2.			
3.			
4.			
5.			

Table 1-3. Measurements of five objects using the metric units of mass, gram, and kilogram, and the corresponding values in pounds.

weight (mass) by this subtraction method when it is not possible to directly weigh an object or substance (like a liquid or a gas).

Your teacher will give you five different objects to weigh. Determine their weight in grams (g) using the balance. Convert their weight to kilograms (kg) and pounds (lbs). Use the fact that 454 grams is equal to one pound or 2.2 lbs is equivalent to a kilogram. Use Table 1-1 to determine the relationship between a gram and a kilogram. Record your results in Table 1-3.

Units of Volume

Volume is the amount of space that a solid object, a liquid, or a gas occupies. The standard U.S. units of liquid volume are the **pint**, **quart**, and **gallon**. Remember that 16 liquid ounces is 1 pint, 2 pints equal 1 quart, and there are 4 quarts in each gallon. The metric units of volume are the **milliliter** (abbreviated ml), **cubic centimeter** (abbreviated cc), and **liter** (abbreviated l). You might need to mix two substances together or dilute a substance by parts (9 parts water and 1 part liquid fertilizer, for instance). Liquid pesticides are prepared for spraying by combining a small amount of the concentrated chemical with a much larger amount of water. When working with liquids, the most frequent measurement is volume. Volumes of liquids measured on the U.S. scale can be converted to the metric scale using this relationship: 1.06 quarts is equal to 1 liter. Cubic centimeters (abbreviated cc) are frequently used in human and animal medicine. Medication for animals may be dosed in cubic centimeters per unit of body weight. One cubic centimeter is equal in volume to 1 milliliter.

Commonly, you will be using measuring cups of 1 cup, 2 cup, and 2-quart capacity to make accurate measurements of liquid volume. The measuring cups may have a metric scale on one side of the cup and a U.S. scale on the opposite side. To determine the volume of a liquid, you need to look for

U.S. Units		Metric Equivalent (wet)	Metric Equivalent (dry)
gallon (gal)	4 quarts	3.785 liters	none
quart (qt)	2 pints	0.946 liter	1.101 liters
pint (pt)	2 cups	0.473 liter	0.550 liter
Metric Units		**U.S. Equivalent (wet)**	**U.S. Equivalent (dry)**
liter (l)	1,000 milliliters	1.057 quarts	0.908 quart
milliliter (ml)	0.001 liter	0.0021 pint	0.0018 pint

Table 1-4. Some units of fluid and dry volume.

the lowest place on the liquid surface when viewed from the side, known as the **meniscus**. The place where the meniscus intersects the scale on the side of the cup is the volume.

Table 1-4 describes some metric and U.S. units of volume. Other conversions that you might use are

$$3 \text{ teaspoons} = 1 \text{ tablespoon}$$
$$2 \text{ tablespoons} = 1 \text{ ounce}$$
$$8 \text{ tablespoons} = 4 \text{ ounces} = \tfrac{1}{2} \text{ cup}$$

1. How many liters is 2 cups of liquid? Show your work in the space provided.

2. Suppose you want to make a 10 percent solution (by volume) of colored water. Your teacher tells you that the final volume of the 10 percent solution is 1 liter. Describe how would you make the solution. How much water would you use to dilute the colored solution? How much of the original solution would you use?

Ingredient	Metric Value	U.S. Value
Sugar	0.72 liter	
Butter (or margarine)	0.24 liter	
Vanilla Extract	5 milliliters	***Hint:*** convert to teaspoons
Flour	1.4195 liters	
Eggs	3 large	3 large

Table 1-5. U.S. equivalents for metric volume values for tea cakes.

3. Describe the color of the original solution and the 10 percent solution. Which liquid is darker in color? Why?

4. You have a friend in France who has sent you a recipe for tea cakes, a type of sweet cookie. You want to make some cookies to bring to school, but you realize the recipe uses metric volumes. Before you can make the cookies, you have to convert the recipe ingredients to U.S. units. Fill in Table 1-5 with the U.S. equivalents for the tea cake ingredients.

Units of Heat: Temperature

The most familiar metric unit to most students is the unit of heat, expressed as degrees Celsius (°C). Many weather stations broadcast agricultural report temperatures in both Fahrenheit and Celsius. Temperatures in °C are commonly displayed on banks and other signs around town. We know that water freezes at 0°C and boils at 100°C. Similarly, water freezes at 32°F and boils at

Temperature °C	Temperature °F
21	
	98.6
	−5
	180
75	

Table 1-6. Equivalent values of temperature on the Celsius (°C) and Fahrenheit (°F) scales.

212°F. What is the relationship between the Celsius and Fahrenheit temperature scales?

$$°C = (°F - 32) \times (5/9)$$
$$°F = [(°C) \times (9/5)] + 32$$

Using the equations above, fill in the missing values in Table 1-6.

Use a thermometer with both Celsius (°C) and Fahrenheit (°F) scales to estimate temperature for the following. (Let the thermometer equilibrate for 10 minutes before recording your temperature readings.)

Temperature in the classroom ____________ °C
Temperature in the classroom window ____________ °F
Temperature in hallway outside of the classroom ____________ °F
Temperature outside ____________ °C

Questions for Thought

1. Why is it important for all scientists to use the same scale of measurement?

2. Can you think of one advantage of using the metric system? Any disadvantages?

3. Which system, the U.S. Customary System or the metric system, do you think will be in use by the year 2050 in the United States? Why?

4. Do you think that the metric system is practical for daily use in agriculture? Explain your answer.

5. Have you used metric units recently (other than today)? Think about temperature, volume, and distance. Where did you see metric units being used?

Glossary

Celsius: abbreviated °C, a temperature scale in which water boils at 100°C and freezes at 0°C.

centimeter: abbreviated cm; a metric measurement of length, 0.01 meter.

cubic centimeter: abbreviated cc; a metric measurement of volume based on a cube with 1-cm sides; also equivalent to 1 milliliter (ml).

Fahrenheit: abbreviated °F; a temperature scale in which water boils at 212°F and freezes at 32°F.

foot: abbreviated ft; a basic unit of length in the U.S. Customary System; derived from the average length of the human foot.

gallon: a measure of liquid volume that equals 231 cubic inches or 4 quarts.

gram: abbreviated g or gm; the basic metric unit of mass; equal to the mass of 1 cubic centimeter (milliliter) of water.

inches: abbreviated in. or "; an inch is a unit of length in use in U.S. Customary System equal to 1⁄36 of a yard or 1⁄12 of 1 foot.

kilogram: abbreviated kg; a metric measure of mass equal to 1,000 grams.

kilometer: abbreviated km, a metric measurement of length, 1,000 meters.

liter: abbreviated l; a basic metric measure of volume, equal to the volume of one kilogram of water at maximum density.

meniscus: relating to liquid; the curved portion of the liquid at the surface, concave when the walls of the cylinder are wetted and convex when they are not. The meniscus is the point at which volume is measured in a graduated cylinder or pipette.

meter: abbreviated m; basic metric measure of length, equal to 39.37 inches.

metric system: a measurement scale based upon multiples of the number 10. The basic units of measurement are the meter, the gram, and the liter.

milliliter: abbreviated ml; a metric unit of volume, 0.001 liter.

millimeter: abbreviated mm; a metric unit of length, 0.001 meter.

ounces: abbreviated oz; an ounce is a unit of mass based on the Roman pound and equal to $\frac{1}{16}$ of 1 U.S. pound.

pint: a unit of liquid volume equal to 28.875 cubic inches.

pound: abbreviated lb; a unit of mass in the U.S. Customary System equal to 16 ounces.

quart: a unit of liquid volume that is equal to ¼ of 1 gallon, also equal to 57.75 cubic inches.

triple-beam balance: a laboratory instrument used to weigh relatively small quantities of a substance.

EXERCISE 2

How Scientists Ask and Answer Questions

Student Objectives

After completing this lab activity, you should be able to:

- List the steps involved in the scientific method.
- Write a question (hypothesis) that would help solve a problem using the scientific method.
- Identify the dependent and independent variables or factors in the statement of a problem.
- Give the purpose of a control in a scientific experiment.

Recommended Resources:

Chapters 1 and 2 in *Exploring Agriscience, 5th Edition.*

Introduction

Most of the facts found in textbooks come from the process of scientific investigation. Scientists study people, behavior, chemicals, weather, and living organisms using the same process to ask and answer questions. This process is called the **scientific method**. Researchers have a high level of curiosity about the world. As a result of their training, they are able to use this curiosity to answer important questions. The scientific method is a way to solve problems in everyday life. As students of agricultural science, you will face many problems in the field and livestock barn. You will need to identify the problem, identify things that might cause the problem, and propose possible solutions. To become good at using this problem-solving method, you must first learn how it works.

Let's explore the scientific method using an example from agriculture. After studying last year's crop yields, a producer in Nebraska decides that he needs to add some additional slow-release nitrogen fertilizer to his soil to improve this year's crop. He calculates how much fertilizer he needs based on the total area of the field. Unfortunately, he calculates the fertilizer required based on the lowest application rate per acre. During fertilizer application, the producer realizes that he will run out of fertilizer after he has fertilized just half of the field at a high application rate. He reduces the amount he puts on the rest of the field. He runs out of fertilizer 50 feet from the edge of the field.

He plants his wheat crop and waits for the seeds to grow. Two months later, the plants in the area of the field that received a high rate of fertilizer are green and healthy. In the part of the field that received less fertilizer, the plants are not as tall and are slightly yellow. Fifty feet from the edge of the field, the plants are very yellow and only 4 inches tall. The grower wonders whether the lower level of nitrogen in the soil could be responsible for the poor growth and yellow color of the young wheat. The county extension office calls you to consult with the grower to try to find a solution to his problem so that he can still produce a good wheat yield from his entire field.

There are several steps involved in the scientific method. The first step is to make a **hypothesis.** A hypothesis is a statement of the problem, and it usually is written as a question. An example of a hypothesis that could be developed from the wheat grower's problem is, "Is the yellow color and poor growth of the wheat related to the amount of nitrogen in the soil from the field?"

A hypothesis should be testable. In the previous hypothesis, two parts or **variables** are given—wheat growth and amount of nitrogen in the soil. If the grower adds nitrogen to the soil in the part of the field where the wheat seedlings are stunted and yellow, can we observe a change in the growth rate and appearance of the wheat? The variable that is changed, the amount of nitrogen in the soil, is the **independent variable**. The growth rate of the wheat plants, which changes with the amount of nitrogen in the soil, is called the **dependent variable**. That is to say, the growth rate of the wheat plants is directly related to the amount of available nitrogen in the soil.

You talk to the wheat producer and use all the information that you collect to make a prediction about the hypothesis. You visit the field and notice poor plant growth in one area. You take soil samples in several areas of the field to see whether there is a difference in soil nitrogen levels. The grower wonders whether adding some additional fertilizer in the places with poor growth would help the wheat grow better.

A next step might be to design an experiment to test the hypothesis (see Figure 2-1). In the experiment, the independent variable is changed, and its effect on the dependent variable is observed. Let's design an experiment to test your question (hypothesis), remembering that the independent variable is the soil nitrogen and the dependent variable is wheat seedling growth.

You collect several bags of soil from the three areas of the field and label them as follows: *large amount of fertilizer*, *small amount of fertilizer*, and *no fertilizer.* You fill two nursery flats with the soil containing a large amount of fertilizer. You fill two more nursery flats with soil containing only a small amount of fertilizer. In two more nursery flats you put soil that has not had any fertilizer added. The final two flats are filled with a commercial greenhouse growing mixture. In each

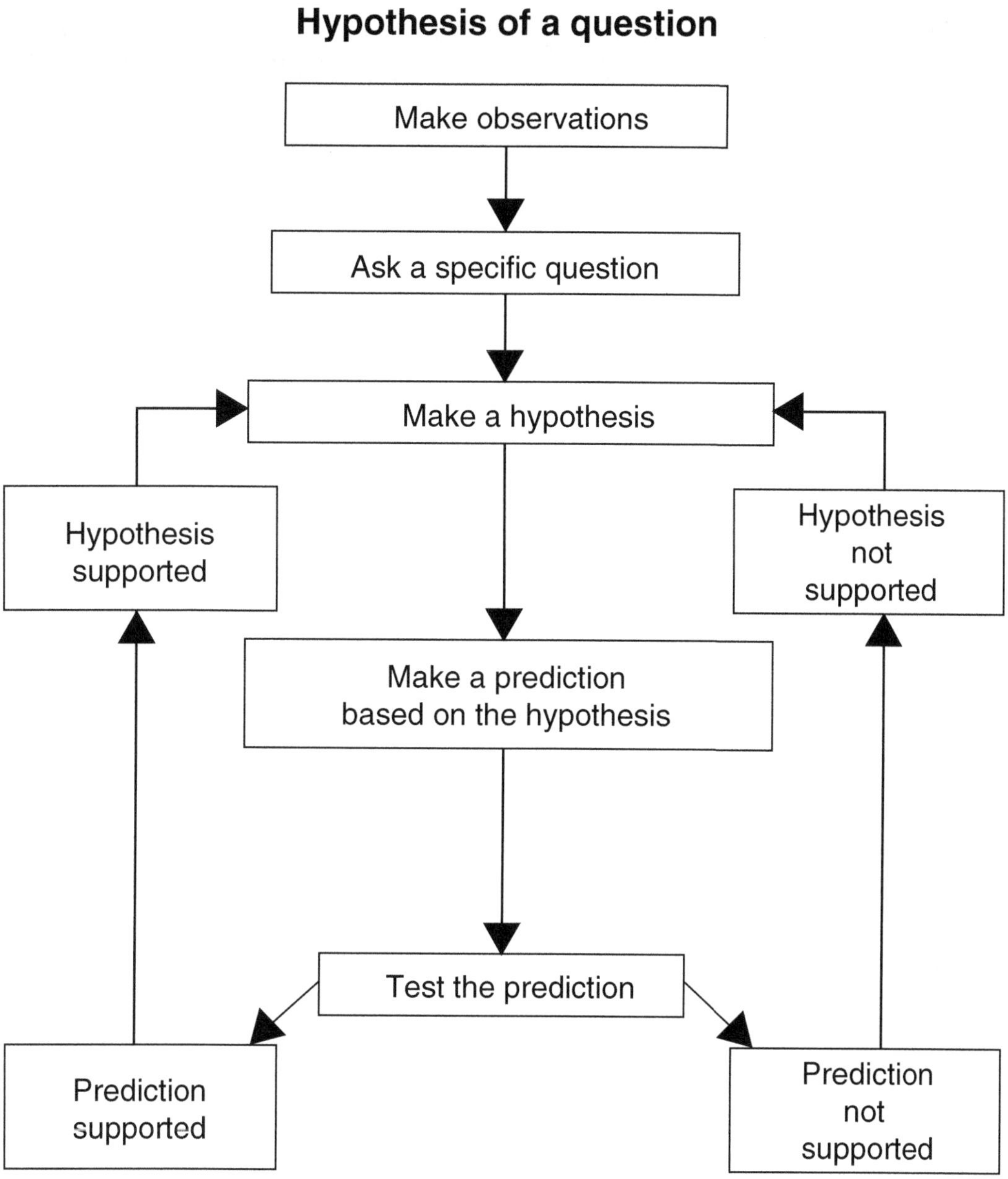

Figure 2-1. The scientific method.

nursery flat 25 seeds (which you obtained from the grower) are planted in five rows spaced 2 inches apart. All eight flats are watered and placed on a mist bench in a greenhouse under grow lights. Each flat receives the same amount of water and light each day, and the temperature is held at 70°F. After eight weeks, you measure each plant in the eight flats every three days for three weeks. You add up the plant heights (in centimeters) for each of the 25 plants and divide the total by 25 to get the average height for all 25 plants in each nursery flat. This step of the scientific method involves making observations and collecting evidence or information. At the end of three weeks, your notebook data will look like Table 2-1.

The numbers in each column represent the average plant height in centimeters. It seems that as the amount of fertilizer is reduced, plant growth decreases.

Flat/Day	1	4	7	10	13	16	18	21
#1 – large amount of fertilizer	4	4.25	4.35	4.8	5.32	5.5	5.81	6.13
#2 – large amount of fertilizer	3.75	3.98	4.25	4.75	5.21	5.62	5.84	6.05
#3 – small amount of fertilizer	3.25	3.31	3.42	3.48	3.55	3.6	3.75	4.1
#4 – small amount of fertilizer	3.41	3.5	3.59	3.62	3.66	3.7	3.79	3.80
#5 – no fertilizer	2.9	2.9	2.98	3.12	3.14	3.25	3.3	3.35
#6 – no fertilizer	1.56	1.6	1.85	2.25	2.43	2.6	2.84	2.97
#7 – commercial growing mixture	4.5	4.75	4.89	5.1	5.15	5.28	5.58	5.85
#8 – commercial growing mixture	5.13	5.25	5.4	5.55	5.7	5.89	6.2	6.33

Table 2-1. Wheat seedling growth and the amount of soil fertilizer.

You share your results with the producer. He asks why you used soil from all three areas of the field, why you planted two flats with each kind of soil, and why you used a commercial potting mixture for two of the flats. You tell the producer that the plants growing in the commercial soil mixture are **controls** for the experiment. You needed to be able to tell if the growth problems were a result of the kind of seeds he used. This is important because the plants could have been growing slowly as a result of bad seeds. Since the seeds in flats #7 and #8 grew as well as the plants in flats #1 and #2, the kind of seeds he used did not have any effect on the growth of the wheat. You conclude that the plants did not grow as well when less fertilizer was added to the soil.

You analyzed your results with respect to the original question. The observations that you made in the greenhouse support your original hypothesis, but the experiment does not provide you or the producer with any solutions. So you suggest to the producer that you look at some additional information that you collected while conducting the experiment. You give him the results of the soil samples that were analyzed at the local college. The soil from the unfertilized part of the field had a nitrogen level of 4.7 mg nitrogen/g soil. In the section of the field that had a small amount of fertilizer added to the soil, the soil nitrogen level was 6.8 mg nitrogen/g soil. The area that received the greatest amount of fertilizer has 9.2 mg nitrogen/g soil. You also tested the commercial potting mixture, which had 9.6 mg nitrogen/g soil. The producer decides that next year he will add fertilizer until the soil nitrogen level is about 9.5 mg nitrogen/g soil.

Practice Problems

Problem #1

You are raising hogs for market, and your veterinarian recommends that you change the kind of feed given to the adult hogs. The veterinarian is concerned that the feed you are using is too high in protein. Although a high-protein diet helps young pigs grow, food too high in protein can cause kidney problems in hogs. You change the feed and notice that the hogs start to lose weight. You want healthy animals that weigh as much as possible, but you do not know how to solve the problem. You call the agriculture teacher for help.

1. Write a hypothesis that describes the problem.

2. Identify the independent variable.

3. Identify the dependent variable.

4. Design an experiment to test the hypothesis you have written.

Problem #2

You are working in a chicken house, and a shipment of young chicks arrives from the hatchery. The chicken house was completely cleaned after the last group of broilers was shipped to the poultry plant two weeks ago. The new chicks appear healthy, but after a week, 25 percent of the chicks die. During this week (the third week of July), it rained hard one evening and the weather conditions remained very warm and humid for the next three days. Your boss thinks that the feed may have gotten wet and gone bad. You think that it is too hot for the chicks and that a better fan might help cool the chicken house. You call the county extension agent for help.

1. Write a hypothesis that describes the problem.

2. Identify the independent variable(s).

3. Identify the dependent variable(s).

4. Design an experiment to test the hypothesis you have written.

Problem #3

You have two small catfish ponds that were recently stocked with fingerlings (small fish about the length of your finger). You notice that a number of the fish in one pond are dying. This pond seems to have some "green stuff" floating in it, and the fish from this pond smell strange when the meat is dressed. You think that the "green stuff" might be killing the fish, but you are not sure. Your agriculture teacher calls a professor from the nearest college to help you.

1. Write a hypothesis that describes the problem.

2. Identify the independent variable(s).

3. Identify the dependent variable(s).

4. Design an experiment to test the hypothesis you have written.

Questions for Thought

1. List the steps in the scientific method.

2. Is the scientific method just for scientists? Why is it important that a non-science student understand the scientific method? How do you think that this method can be used in everyday life?

3. Describe the difference between independent and dependent variables. Give an example of each.

4. Why is a control used in an experiment?

Glossary

control: the subjects in an experiment (plants, animals, or people) that do not receive the treatment; all variables are constant. Results from the groups that receive the experimental treatment are compared for each variable with the observations from the control group to determine whether differences are a result of the experimental "treatment."

dependent variable: the factor that is measured in the experiment and is affected during the experiment.

hypothesis: a guess that is based on earlier observation(s) and is the starting point for more scientific thought and/or experiments.

independent variable: the factor that is changed or controlled in an experiment; its effect on the dependent variable is observed.

scientific method: a systematic way of inspecting, recognizing, and stating scientific problems; designing experiments, making observations, and looking at the results of experiments to explain and simplify one's understanding of the problem.

variable: something in an experiment that can have more than one value; for example, temperature is a variable if observations are made at different temperatures such as 0°C, 10°C, 25°C, and so on.

EXERCISE

3

Tools of Science—Microscopes

Student Objectives

After completing this lab activity, you should be able to:

- Name the parts of the compound microscope and give their function.
- Name the parts of the dissecting (stereo) microscope and give their function.
- Distinguish between parfocal and parcentric microscopes.
- Calculate the total magnifying power of a microscope given the magnification of the ocular and objective lenses.
- Determine when to use a compound microscope and when to use a dissecting microscope to get the best view of something.

Recommended Resources:

Chapter 1 in *Exploring Agriscience, 5th Edition.*

NOTE:

If you wear glasses, take your glasses off before using the microscope. The lenses of the microscope will be adjusted according your visual acuity when you focus the microscope.

Introduction

All scientists use special tools to make observations. Depending upon the kind of science, scientists use many different kinds of machines and tools. Scientists use balances to weigh chemicals or living organisms, special glassware for chemical reactions, instruments that tell the scientist what chemicals are in an unknown substance, computers to store information and analyze results, or specialized microscopes to view objects. Every scientist makes use of many different tools, and as students of agriscience, you will learn to use many of them. In this lab activity, you will learn about two different kinds of microscopes, and when to use each kind.

Early biologists could not observe objects smaller than what the human eye could see accurately. They could only observe organisms and nonliving things that were 1 mm (0.394 in) in size or larger (see Figure 3-1). To investigate the structure of plants and animals more closely, some type of magnification was needed. In 1677, a Dutch scientist named Anton von Leeuwenhoek (pronounced LOO-ven-hook) made the first simple **light microscope**. With his invention, Leeuwenhoek looked at pond water, human blood cells, and other samples. He was amazed by the small moving organisms he saw in each of them.

There were no improvements on Leeuwenhoek's invention until the English scientist Robert Hooke made a device similar to Leeuwenhoek's to look at cork. Cork is made of dead plant cells; only the walls around the dead cells remain. Hooke looked at the dead cells arranged in a regular pattern and thought that they resembled tiny rooms, or "cells."

As the study of living organisms continued, the light microscope became more powerful. With the light microscopes currently in use, scientists can look at objects that are as small as 1 μm (10^{-6} m or 0.00003937 in) (as shown in Figure 3-1). To look at specimens that are even smaller, scientists use electron microscopes. They use an electron beam instead of a light beam to illuminate the object being studied.

Light Microscopes

The type of microscope used in most biology courses is the light microscope. The light microscope allows the observer to look at living cells and cell activities. There are two kinds of light microscopes: the **compound microscope** and the **dissecting microscope** (also called a **stereo microscope**). The difference between these is the **magnification** and **resolution** of each microscope. Magnification is the amount of increase in an object's size. Resolution is the smallest distance between two points at which the two points are both visible. The compound microscope has greater magnifying power and better resolution than the dissecting (stereo) microscope.

Compound Microscope

The compound microscope has at least two lenses, the **ocular** and the **objective** lenses (see Figure 3-2). The ocular is the lens at the top or front of the microscope where you place your eye. Microscopes that have two eyepieces are called binocular microscopes; monocular microscopes have just one eyepiece. The objective lens is at the end of the body tube. The body has reflective mirrors inside, similar to those of a telescope. Most compound microscopes have several objective lenses of varying magnification. The shortest lens is the low-power objective lens and usually has a magnification of 4 × or 10 × depending upon the microscope. Microscopes that lack a condenser usually have a high-power objective lens that can magnify things 40 ×. The total magnification of an object is greatest when the high-power objective lens is in use.

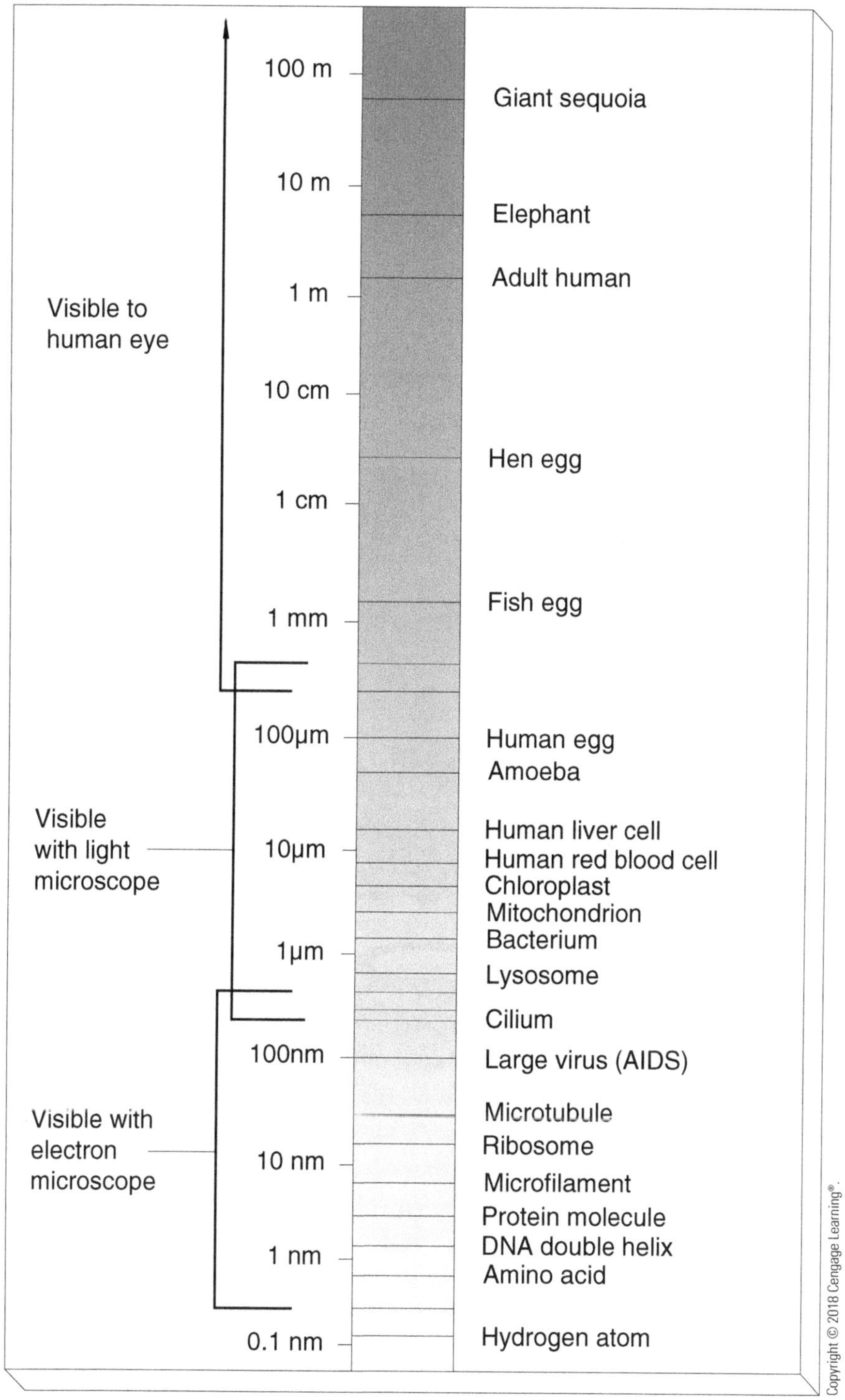

Figure 3-1. The sizes of many different living things.

The **condenser with an iris diaphragm** is found on some compound light microscopes. The condenser and iris diaphragm direct light evenly at all parts of the object on the microscope slide. The condenser increases the resolving power of a microscope. Microscopes that use oil immersion objective lenses also have a condenser with an iris diaphragm. The condenser has an adjustment knob that

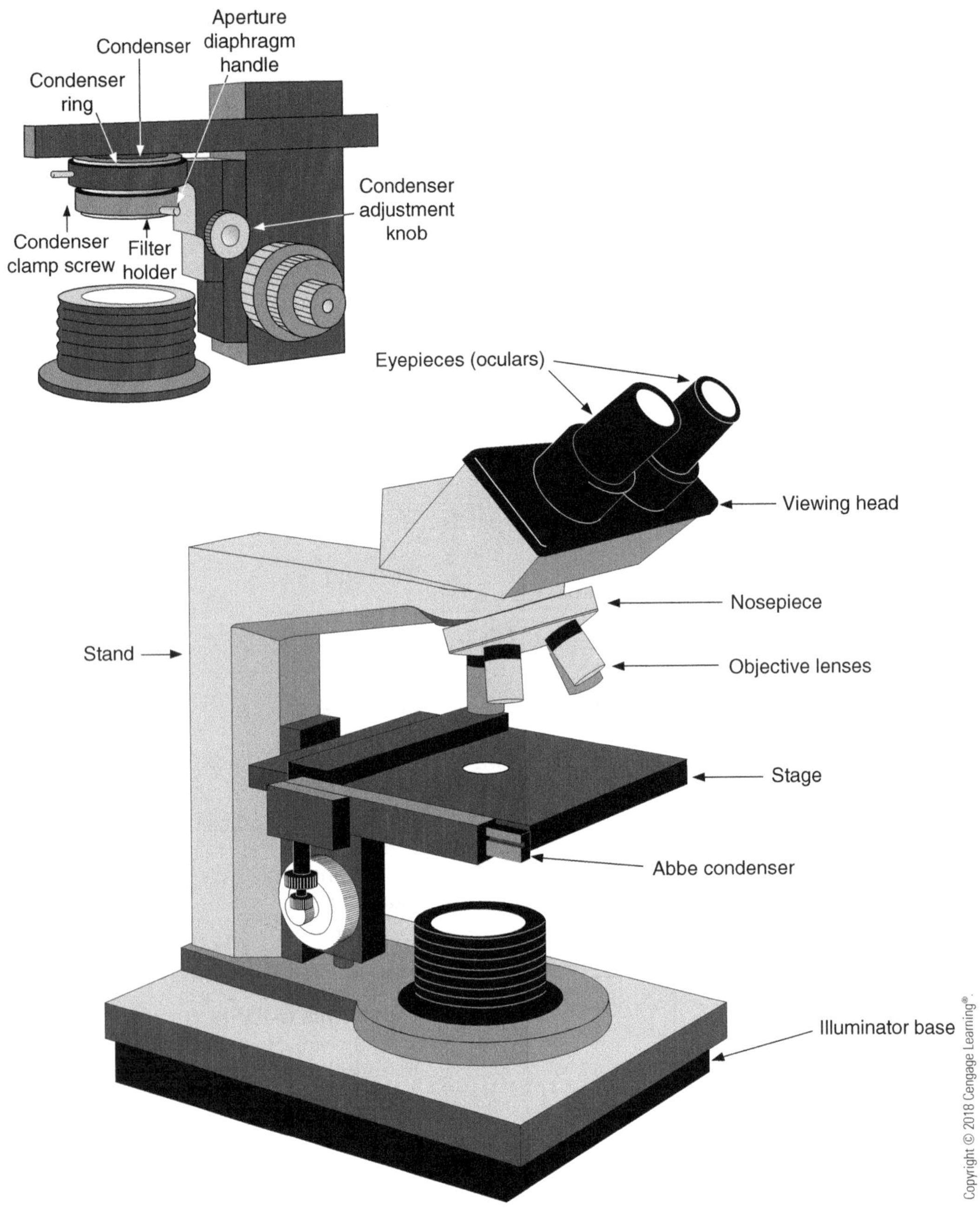

Figure 3-2. The parts of a compound light microscope.

allows you to change the distance between the condenser and the bottom of the microscope slide. When you turn the **condenser adjustment knob**, the condenser moves closer to or farther away from the slide.

The iris diaphragm is like the part of a camera that lets light into the camera box. By opening and closing the diaphragm, you can adjust the amount of light that passes through the object on the microscope slide.

The highest-power objective lens, the oil immersion lens (usually 100 × magnification), is used by placing a drop of special oil (called immersion oil) on the slide, adjusting the condenser so that it is as close to the microscope slide as possible, and adjusting the iris diaphragm to allow the most light to pass through the slide. The oil immersion lens is used to increase the clarity of the details on the side or to view very tiny organisms such as bacteria.

The formula to determine the total magnifying power of a microscope is:

$$\text{Ocular} \times \text{Objective} = \text{Total Magnification}$$

Activity: Observing Magnification

Materials Needed

- A calculator
- Slide with letter "e"
- Slide with three different colored threads

Look at your microscope. Calculate the total magnification for each objective lens by completing Table 3-1.

Obtain a slide of the letter "e" from your teacher. Put the slide on the microscope stage so that the letter "e" appears right-side-up (correctly placed for reading). Put the slide in the center of the **stage** (flat table portion under the revolving nosepiece). Try to adjust the slide so that the light coming through the stage brightens your letter "e." This means that the specimen is properly centered on the stage. Turn the nosepiece so that the low-power objective lens is in place. While looking through the eyepiece(s), turn the **coarse adjustment knob** (the largest knob on the side of the microscope) until your letter "e" comes into view. Make your best focus with the coarse adjustment knob, then use the **fine adjustment knob** (the smaller knob on the side—see Figure 3-2) to make the final focus.

a. When looking through the microscope, how does the appearance of the letter "e" compare with the orientation on the slide?

b. Move the slide to the right while looking through the ocular(s). Which direction does the slide move? With any compound light microscope, the lenses produce an image (what you see through the microscope) that is upside down and backwards from the way it is on the slide. If you are looking through the microscope and wish to move the image to the right, you will need to move the slide to the left. It will take some practice to get used to this.

Ocular	× Objective	= Total Magnification
OCULAR ___	LOW-POWER OBJECTIVE ___	=
OCULAR ___	MEDIUM-POWER OBJECTIVE ___	=
OCULAR ___	HIGH-POWER (DRY) OBJECTIVE ___	=
OCULAR ___	OIL IMMERSION OBJECTIVE ___	=

Table 3-1. Magnifying power of the compound microscope.

Now that you are in focus on lower power, move the revolving nosepiece so that the next objective lens is in place. You should find that the "e" is still in focus. If properly adjusted, the compound light microscope is **parfocal**. That means that when an object is in focus using the low-power objective lens and the revolving nosepiece is moved to a higher-power objective, the object should still be in focus. Also, the compound light microscope is **parcentric**. If your letter "e" is in the center of your field of view (the circle of light you see when you look through the ocular lenses), it should remain centered when you change to another objective lens.

Depth of field is another characteristic of a light microscope. This is the working distance between the objective lens and the specimen. Compound and dissecting microscopes have differing depths of field. The greater working distance is found in the dissecting microscope.

The microscope slide is three-dimensional. To experience depth of field, you will need to get a slide with three different colored threads on it. These threads are crossed at different places on the slide, and the slide is fairly thick. You will need to use extra caution in using this slide with the longer (higher magnifying power) objective lenses. Start by placing the colored slide on the stage so that the place where the threads cross is in the middle of the circle of light. Using the low-power objective and the coarse adjustment knob, bring the colored threads into focus. Make any necessary adjustments in the position of the threads on the stage. Move the revolving nosepiece until the high-power objective is in place. Using the fine adjustment knob, bring the colored threads into focus. *Remember, the coarse adjustment knob is not used with the high-power or oil immersion lenses!* While looking through the ocular and moving the fine adjustment knob, you should see several different thread colors come in and out of focus. This illustrates depth of field.

a. Which thread color is on top?

b. Which is in the middle?

c. Which is on the bottom?

Total Magnification	Width of Field of View
(low-power)	
(medium-power)	
(high-power, dry)	

Table 3-2. Relationship between the total magnification and the width of the field of view in a compound light microscope.

There is a reverse relationship between the size of the field of view and total magnification. That is, as magnification increases, the size of the field of view (the circle of light you see when you look through the oculars) decreases. To test this relationship, obtain a clear plastic metric ruler from your teacher and clip it down on the stage using the stage clips. Try to center the metric scale in the middle of the field of view. Focus on the ruler using the low-power objective lens. How many millimeters (mm) wide is the field of view? Record your measurement in Table 3-2. (**HINT**: Total Magnification = Ocular × Objective.)

Switch to the next objective lens and focus again on the metric scale on the edge of the ruler.

a. How many millimeters wide is the field of view? Record your measurement in Table 3-2.

b. Has the field size increased or decreased?

c. Switch to the high-power objective lens. Are any divisions on the edge of the metric scale visible?

d. How wide is the field of view? Record your measurement in Table 3-2.

The inverse relationship between field size and magnification frustrates many new microscope users. "I've lost it!" is frequently heard. In fact, nothing is lost. The specimen was not in the center of the field of view. As the field size decreased as the magnification increased, the object was no longer in the field. A quick adjustment of the slide on the microscope stage can usually help you find the lost item.

Activity: Making a Wet Mount

Materials Needed

- Lens paper
- Clean slide
- Clean coverslip
- Sample of pond water (or other living culture)
- Samples of three organisms to examine under the dissecting microscope

To look at live organisms with a microscope, the organism or specimen must be protected from the heat created by the microscope light; the specimen is usually in water or some other liquid, called a wet mount. Get a clean slide and coverslip (sometimes called a coverglass) from your teacher. Clean each eyepiece (ocular) and all the objective lenses with lens paper before you begin and after you have finished. This will keep your microscope in good working order. Place the slide on a flat surface and add a small drop of pond water or some other living culture on the center of the slide. Take the coverslip and place on edge so that it touches the drop of liquid at an angle (see Figure 3-3). When the edge of the drop of liquid touches one side of the coverslip, drop the coverslip onto the liquid. It should flatten out with very few air bubbles. Place the slide on the stage and begin by focusing using the low-power objective. If you see any clear circular objects outlined in black, these are air bubbles. Gently tapping the surface of the slide with a pencil eraser should force any air bubbles out the edge of the coverslip. After your slide is in focus under low power, switch to the next objective, using the fine adjustment knob to make your focus. Draw what you see in the space provided.

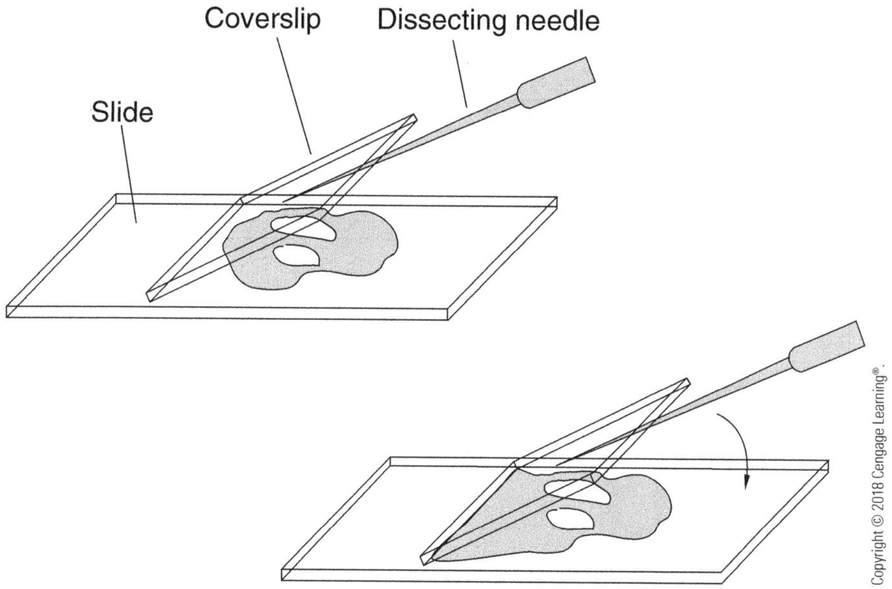

Figure 3-3. How to make a wet mount on a microscope slide.

Dissecting (Stereo) Microscope

The dissecting microscope (also called a stereo microscope) was developed for use in dissecting small specimens. The dissecting microscope (see Figure 3-4) has a greater working distance and depth of field than the compound light microscope (as shown in Figure 3-2). The light can shine on the specimen from the bottom (as with the compound microscope) or from the top.

The total magnifying power of the dissecting microscope is far less than that of the compound light microscope. As a result, dissecting microscopes are used to view surface details of larger objects, dissect small specimens (such as chicken eggs, insects, or flowers), or to view larger living specimens such as the flatworm, earthworm, or leech.

Your teacher has a variety of different objects, living and nonliving, for you to use with the dissecting microscope. Draw and label three different specimens in the space provided.

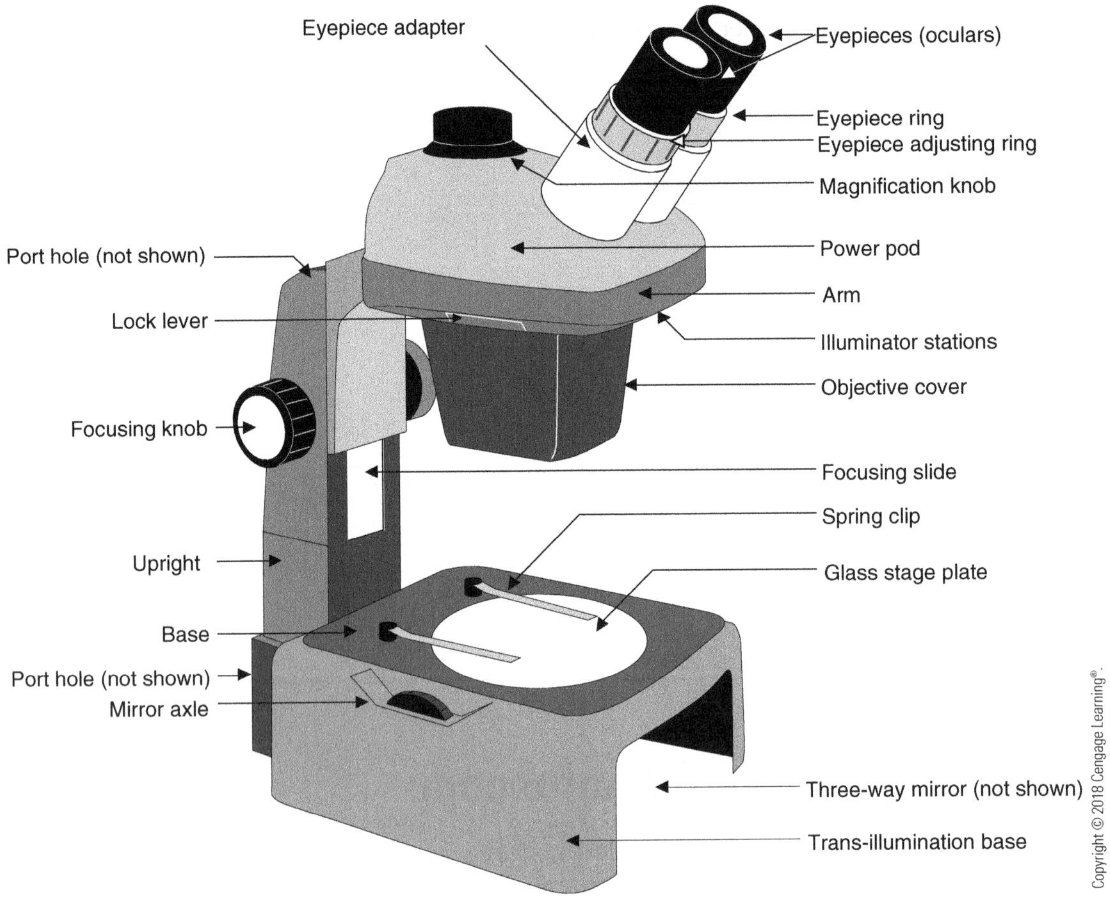

Figure 3-4. The parts of a stereo (dissecting) microscope.

Questions for Thought

1. What is the difference between parfocal and parcentric? How do the parfocal and parcentric properties of the compound light microscope affect its use?

2. How is the total magnifying power of a microscope calculated?

3. Under what circumstances would you use a compound light microscope?

4. Under what circumstances would you use a dissecting microscope?

5. Describe the relationship between field of view size and total magnification. How does this affect microscope use?

6. Why is the microscope such an important scientific tool? How different do you think life would be without the microscope?

Glossary

coarse adjustment knob: large outside adjustment knobs located on each side of the microscope at the base (see Figure 3-2); used to make the beginning focus of the slide; *the coarse adjustment knob is used with the 4 × and 10 × objectives only.*

compound microscope: a microscope that has both and ocular and objective lenses mounted in a body tube; uses light to illuminate the viewing specimen; specimen is illuminated from the bottom; with powerful objective lenses, very small specimens (such as bacteria and sperm) can be viewed (see Figure 3-2).

condenser adjustment knob: located below the stage on the left side of the microscope; regulates the distance of the condenser from the stage and assists in focusing the light source on the specimen through the condenser lens (see Figure 3-2).

condenser with an iris diaphragm: portion of the microscope that regulates the amount of light passing through the specimen (see Figure 3-2).

dissecting microscope: also called a stereo microscope (see Figure 3-4). A microscope with a larger working distance between the object being viewed and the objective lens. Practically, this means that the object can be illuminated from above and below. The total magnifying power of the dissecting microscope is less than that of the compound microscope, and its use is limited to viewing details of larger objects.

fine adjustment knob: small inside adjustment knob located on each side of the microscope at the base (see Figure 3-2); used to make final focus of the specimen. *The fine adjustment knob is the only adjustment knob to be used with the 40 × and 100 × objectives.*

light microscope: a category of microscope that allows the observer to look at living cells and activities. There are two kinds of light microscopes: the compound microscope and the dissecting microscope.

magnification: the appearance of enlargement by a microscope; the total magnifying power of a microscope is calculated by multiplying the magnifying power of the eyepiece by the magnifying power of the objective lens.

objective: attached to the revolving nosepiece, the objectives contain lenses of differing magnification and are used with the ocular lenses to view the specimen (see Figure 3-2). Most microscopes have several different objectives: 4 × (low-power objective); 10 × (medium-power objective); 40× (high-power objective); and 100 × (oil immersion objective).

ocular: eyepiece(s) containing a lens located at the top or front of the microscope and used to magnify the specimen (see Figures 3-2 and 3-4). Microscopes with a single eyepiece are monocular microscopes, and microscopes with two oculars are binocular microscopes.

parcentric: the property of compound light microscopes in which an object that is in the center of the field of view when using the lower-power objective remains in the center of the field of view when a higher-power objective lens is in place.

parfocal: in practice, a parfocal microscope allows the user to focus on a slide using the 10× objective, and when the user switches to the 40× objective, the slide should still be in focus.

resolution: the smallest distance between two points that can still be viewed separately.

stage: the flat horizontal portion of the microscope on which the specimen is placed for viewing (see Figure 3-2).

stereo microscope: another name for a dissecting microscope. A microscope with a larger working distance between the object being viewed and the objective lens. Practically, this means that the object can be illuminated from above and below. The total magnifying power of the dissecting microscope is less than that of the compound microscope, and its use is limited to viewing details of larger objects.

EXERCISE 4

Physical and Biological Properties of Soil

Student Objectives

After completing this lab activity, you should be able to:

- Tell the difference between organic and inorganic parts of soils.
- Measure the pH of the soil and relate pH to the kind of rocks the soil came from.
- Tell the difference between gravel, sand, loam, silt, and clay, and associate these terms with particle sizes in a soil sample.
- Relate the water-holding ability of a soil to the size of the particles in the soil.
- Identify the part of the soil profile that has the largest number of microorganisms.
- Describe what algae, bacteria, and fungi do in soils.
- Tell the difference between algal, bacterial, and fungal cells using a compound light microscope.
- Talk about the effects of agricultural practices on soil microbes.
- Look at *Rhizobium* in the root nodules of a legume using a light microscope.
- Investigate the effects of pea inoculum on soybean seedling growth.

Recommended Resources:

Chapter 3 in *Exploring Agriscience, 5th Edition.*

Introduction

Plants could not grow without soil. Soils are complicated structures. They result from weather, volcanoes, sediment deposits of rivers and streams, plants, and animals (including humans). All soils have two parts: **organic** and **inorganic** material. Organic materials contain both of the chemical elements carbon (C) and hydrogen (H). Organic material is the product of the activities of living things. When plants, animals, and microorganisms die, their remains become part of the soil. Organic material is usually found near the surface of the soil. Between 3 and 8 percent of the **topsoil** is organic material. Fresh organic matter is ideal for the growth of microorganisms. Soil microorganisms break down fresh organic matter into a substance called **humus**. In the **subsoil**, organic matter is less than 1 to 2 percent of the soil. The mineral part of the soil is inorganic. It lacks the chemical elements carbon and hydrogen and comes from the bedrock, the rock layer below the soil. The physical and chemical characteristics of the soil layers produce soils of different types and determine the suitability of a specific soil for agricultural use.

One way that soil is classified is by texture, the fraction of mineral particles of different sizes. There are four different sizes of particles in the United States Department of Agriculture (USDA) system of soil classification: **gravel**, **sand**, **silt**, and **clay**. Gravel is the mineral part, with a particle size larger than 2.00 mm in diameter. Sand particles are in a range of 2.00 to 0.05 mm in diameter. Silt particle sizes range between 0.05 and 0.002 mm in diameter. Clay particles are smaller than 0.002 mm in diameter. **Loam**, a mixture of different sized particles, is made up of less than 52 percent sand, 28 to 50 percent silt, and 7 to 27 percent clay (see Figure 4-1). Loam is the best soil for plant growth.

Soil pH

Soil **pH** is a measure of the concentration of a charged hydrogen atom (H^+). Soils containing a lot of hydrogen ions are acidic. Soils containing few hydrogen ions are alkaline (or basic). Soil pH is important because the pH of the soil influences plant growth. In acidic soils that contain metals such as iron and aluminum, the metals are more easily absorbed by the roots of plants. At some point, these metal ions become toxic to the plants, and the plant will die. The soil pH of an extremely acidic soil must be raised before plants will grow in it. Soil pH is related to the amount of organic material in the soil and to the pH of the rock from which the soil comes. For example, limestone produces alkaline soils in comparison to granite, which produces acid soils.

Knowing soil pH is important because soil pH influences how well plants grow. The method that you will use to determine pH involves measuring the pH of soil suspended in distilled water. Your teacher will give you two soil samples: Sample A (cultivated field with nitrogen fertilizer added recently) and Sample B (forest soil).

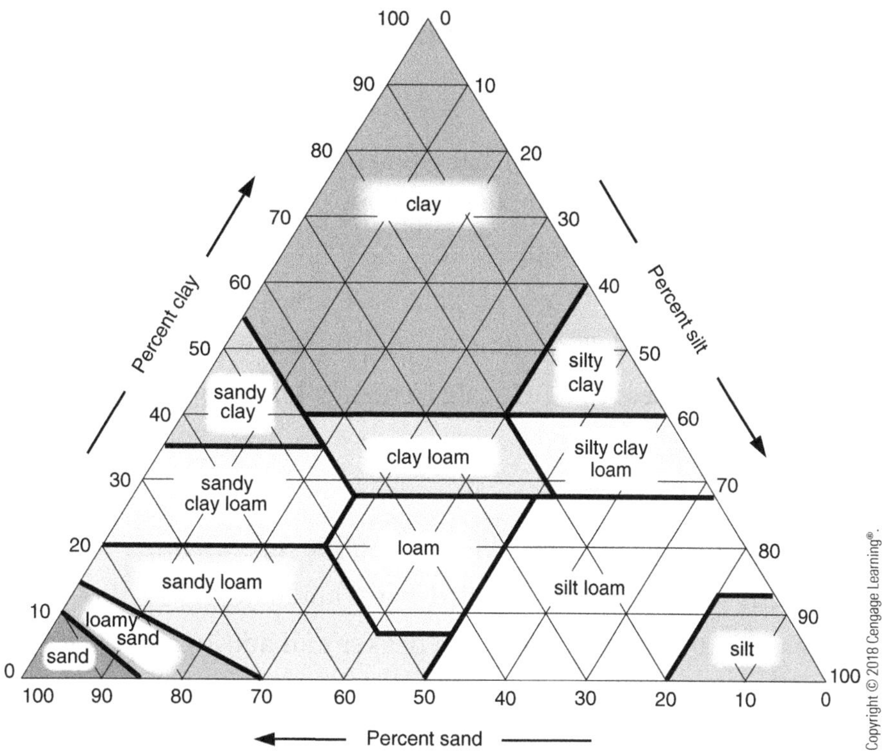

Figure 4-1. Soil particle size and soil texture class.

Questions for Thought

1. What effect do you think cultivation could have on soil pH? (**HINT:** Think about the amount of organic material and chemicals added to the soil during farming.)

2. Which area do you think will have more acidic soil? Why?

3. Which area do you think will have more alkaline soil? Why?

Activity: Measuring Soil Water pH

Materials Needed

- Air-dried soil samples from two sites: one from a cultivated field with nitrogen fertilizer added recently (Sample A) and one from forest soil (Sample B)
- Scale for weighing soil samples
- 25 ml beaker
- Distilled water (20 mL for adding to samples and extra for rinsing beakers)
- pH paper (and access to a pH scale to read color change)

Procedure

1. Weigh out 10 g of air-dried soil from Sample A.
2. Rinse a 25 ml beaker with distilled water and dry.
3. Place your 10 g soil sample in the beaker and add 10 ml of distilled water to the soil.
4. Mix your soil and water in the beaker for 5 seconds.
5. Let the mixture stand for 10 minutes. (***Note:*** Clay soils may require more time to settle.)
6. Gently swirl the soil–water mixture, and be certain that your pH paper is just above the soil particles in the beaker.
7. Allow the pH paper to remain in the soil–water mixture for about 1 minute.
8. Remove the pH paper and compare the color of the paper to the pH color scale that your teacher gives you.
9. Write the soil pH for Sample A in Table 4-1.
10. Repeat steps 1 through 8 with a clean beaker and soil from Sample B. Write the soil pH for Sample B in Table 4-1.
11. Compare the soil pH for Sample A and Sample B. Are the results what you expected?

Sample/Site	Sample A Cultivated Field With Fertilizer	Sample B Forest Soil
Sample pH		

Table 4-1. Results of soil–water pH for soil samples.

Questions for Thought

1. Think about all that you have learned about soils. Do you think soil pH is connected to the kinds of plants growing in a field or forest?

2. Do you think that the pH of soil is influenced by agricultural practices such as tilling and the application of agricultural chemicals? Explain your answer.

The Amount of Water in Different Soils

Water is essential for all plant growth. Animals must have water to live. Many animals get their water from plants that store water in their leaves, stems, roots, fruits, and seeds. How does soil type relate to the amount of water the soil can hold? Is it possible for soil to hold too much water?

Soil is made of smaller parts called **peds**. Peds determine the structure of soil (in contrast to soil texture, which is related to particle size). Factors that contribute to the soil structure are soil texture (relative proportion of different sized particles); the amount of organic matter; the mineral content of the soil (such as limestone or iron); temperature; activities of earthworms, fungi, and bacteria in soil; and soil cultivation practices.

All these things contribute to the size and shape of soil clumps. Soil has spaces between the peds (clumps) through which water and gases from the air can move. When these spaces are completely filled with water, the soil is saturated. Eventually, some of this water is replaced by gases from the atmosphere such as oxygen, carbon dioxide, and nitrogen. Most plants do not grow well in soils that do not drain. Soils that do not drain water become permanently saturated.

Waterlogged soils typically are low in oxygen. Only plants and microorganisms that do not need very much oxygen grow in waterlogged soils. Swamps, bogs, and tidal marshes are examples of areas with permanently saturated soils. These locations are undesirable for the cultivation of agricultural crops. The amount of water that soils can hold is related to ped size, soil texture, and **slope** of the site. Sites with moderate to steep slopes drain water faster than flat areas. Using soil samples from a cultivated field and a forest, you will estimate the amount of water in a saturated sample.

Questions for Thought

1. Do you think that you will have to add more water to the forest soil or to the field soil to saturate the soil? Why?

Activity: Measuring Soil Saturation

Materials Needed

- Soil samples from two sites: one from a cultivated field with nitrogen fertilizer added recently (Sample A) and one from forest soil (Sample B)
- Two canvas bags for soil samples
- A pan for soaking soil samples in or access to running water
- Scale for weighing soil samples

Procedure

1. Measure 100 ml of soil from each of the two sites: A (field) and B (forest).
2. Put each of the samples in a canvas bag, tie the bag tightly, and weigh the sample. Record the dry weight in Table 4-2.
3. Thoroughly saturate each bag with water by placing it underneath running water in the sink or putting the bag in a pan of water and allowing the bag to absorb as much as it can (approximately 20 to 30 minutes).
4. When the samples are saturated, allow the excess water to drain.
5. Weigh each of the samples again and record the wet weight in Table 4-2.
6. Calculate the weight of the water by subtracting the dry weight from the wet weight, and record the amount in Table 4-2.

Sample Site	Wet Weight	Dry Weight	Water (g) = Wet Weight − Dry Weight
Sample A cultivated field with fertilizer			
Sample B forest soil			

Table 4-2. The amount of water that two different kinds of soil can hold.

Questions for Thought

1. Which soil sample, the field soil or the forest soil, held more water?

2. Is this what you predicted?

3. How do you think the kinds and numbers of plants growing in an area affect the amount of water that the soil can hold?

4. Given the measurements of soil water pH and water saturation for each of the two sites, and assuming that all other factors that affect plant growth such as orientation, slope, and climate are the same, which of the two sites is better suited for growing crop plants? Include in your answer all the evidences that you have gathered in the experiments on physical soil properties.

5. Has the kind of plants growing in each area affected the physical properties of the soils? How?

6. In which area would you expect to find the greater number of different kinds of living organisms in the soil? Why?

7. What advice would you give to growers who are interested in maintaining healthy soil based upon these experiments?

Soil Organisms

In a soil profile, the **A Horizon** contains the organic portion of the soil and the living soil organisms. All soils contain some organic material that is produced by the activities of living organisms. Soil organisms play an important role in recycling organic material. Larger soil organisms, such as earthworms and insects, contribute to the breakdown of dead plant and animal material. Earthworms eat the fungus and bacteria that grow on dead leaves. By consuming the leaves, the earthworms break down the leaf material into smaller particles. The dead material is called **detritus**. The smaller particle size makes the leaf material more attractive for the growth of microorganisms such as bacteria and fungi. The bacteria and fungi chemically break down large organic molecules into smaller molecules. The smaller molecules are more readily absorbed by plant roots.

Some bacteria associated with the roots of leguminous plants and certain cyanobacteria (blue-green algae) are also responsible for increasing soil nitrogen content through a process called **nitrogen fixation**. Nitrogen is essential for plant growth largely because the proteins and nucleic acids (DNA and RNA) necessary for normal cell functioning contain nitrogen. Following nitrogen fixation, nitrogen is present in the soil in the form of ammonia. While ammonia is toxic to some plants and animals, some soil bacteria convert the toxic ammonia to nitrate. Nitrate is the form of nitrogen that is absorbed by most plants.

Many soil bacteria are also involved in the process of decay, which releases organic and inorganic nutrients from dead material into the soil. Some soil bacteria and fungi establish special relationships with plants. For instance, many types of **legumes** produce higher yields when bacteria are present in their roots (see Figure 4-2). Some fungi form **mycorrhizal** relationships with plants and assist them in the uptake of water and nutrients. Mycorrhizal fungi improve plant growth, and there are many examples of economically important mycorrhizal fungi, including truffles. Soil microorganisms provide clues to the quality of a soil. Healthy soils contain large amounts of soil bacteria, fungi, and algae.

Observing Soil Microorganisms

You will get a soil–sterile water suspension from your teacher to look at under the light microscope. Your teacher will provide you with two soil samples: Sample A from a field under cultivation with leguminous plants, and Sample B from a pasture currently in use.

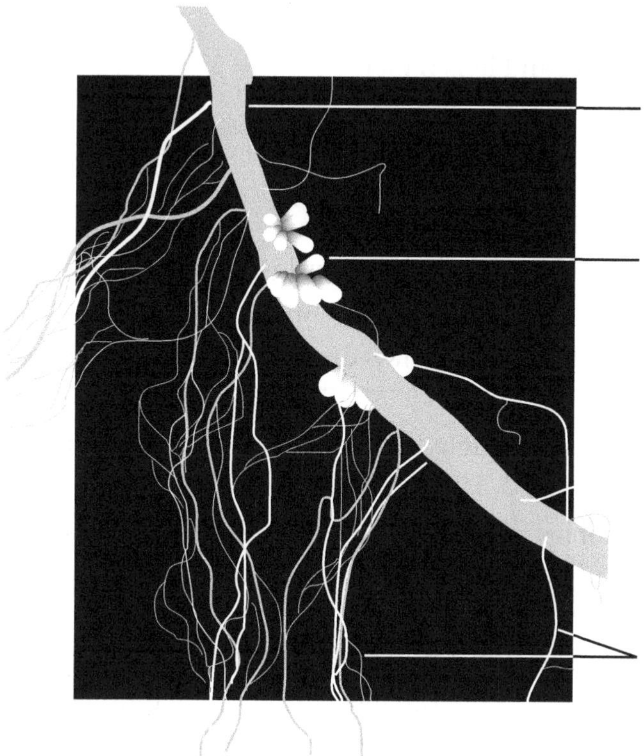

Figure 4-2. Root nodules on a legume.

Questions for Thought

1. How are soil bacteria, fungi, and algae related to the type of cultivation and vegetation in each sample area?

Activity: Observing Soil Organisms

Materials Needed

- Samples from soil–sterile water solutions from a field under cultivation with leguminous plants (Sample A) and from a pasture currently under cultivation (Sample B)
- Compound light microscope
- Clean slides and coverslips
- Sterile pipette

- Immersion oil
- Lens cleaning solution and lens paper
- Crystal violet stain

Procedure

1. With the cap secured, shake the soil–sterile water suspension from Sample A. Allow the large particles to fall to the bottom of the jar.
2. Using a sterile pipette, pipette a few drops of the soil–water suspension onto the middle of a flat slide.
3. With the coverslip at an angle, drop the coverslip on the soil–water suspension. (**HINT:** See Figure 3-3 if you are having difficulty.)
4. If you have air bubbles under your coverslip, gently tap the surface of the coverslip with your pencil eraser or the blunt end of a dissecting needle. The bubbles should move to the outside of the coverslip.
5. Turn on your microscope light and make any necessary adjustments (center the slide on the stage, adjust the light using the diaphragm, etc.). If you are having difficulty with your microscope, review the use of a compound light microscope in Exercise 3.
6. Begin with the low-power objective and bring the slide into focus. Bacterial cells are too small to be seen with the low-power objective, but fungi and some algae may be visible.
7. Without moving the slide, switch to the medium- and high-power objectives. *DO NOT use the coarse adjustment knob with the high-power objective!*
8. Algal cells will appear rectangular or circular with a definite outside wall. The cells of soil algae are usually green, blue-green (very small-celled blue-green algae are also called cyanobacteria), or a brownish color (diatoms). Do you see any algae in the soil–water suspension from sample A? If so, draw what you see in the space provided.

9. Fungal cells are clear in color, appearing as long thin threads, called **hyphae**, with definite walls between the cells. Some of the hyphae may have broken when you shook your sample jar. Do you see any fungal cells in the soil–water suspension from Sample A? If so draw what you see in the space provided.

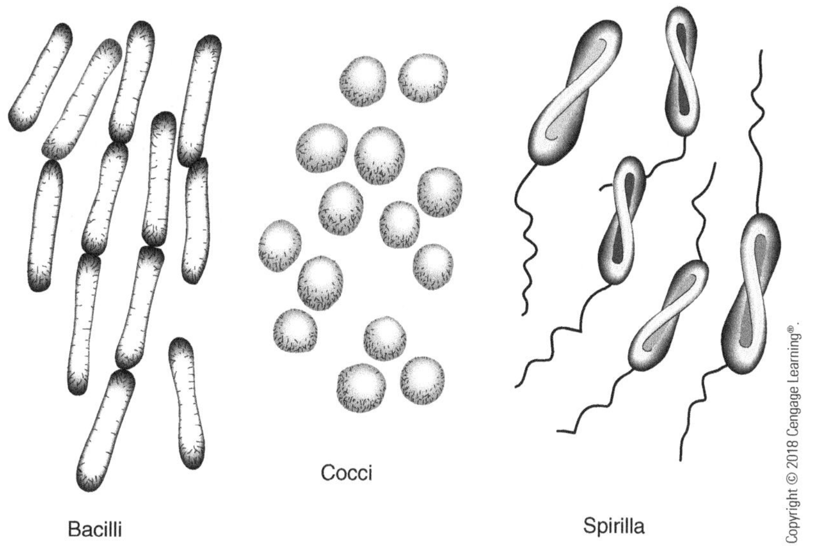

Figure 4-3. Shapes of bacterial cells.

10. Bacterial cells are very small (150 to 500 μm). Many bacteria are too small for inexperienced eyes to see without the use of the oil immersion objective lens. Your teacher will review the use of the oil immersion objective lens with you. Be certain that the slide is in focus using the high-power objective lens, and adjust your condenser so that it is closest to the stage. Increase the amount of light by opening the condenser diaphragm. Move the revolving nosepiece in the direction of the oil immersion lens so that the slide is halfway between the high-power objective and the oil immersion objective. *Do not put the oil immersion objective in place yet.* Put 1 to 2 drops of immersion oil on your coverslip and slide the oil immersion lens in place. Wait a minute or two to assure that the oil has covered the oil immersion objective. *Using the fine adjustment knob only*, bring the slide into focus. Bacterial cells come in three shapes (see Figure 4-3): spherical (**coccus**), rod-shaped € **bacillus**), and spiral (**spirilla**). They are single-celled organisms with a pronounced cell wall and little color. Do you see bacterial cells in your soil–water suspension from Sample A? If you see any bacteria cells, label the cells according to shape.

11. Bacterial cells are easier to see with the aid of special colorizing agents or stains. Move the oil immersion lens out of place. Carefully remove the immersion oil from the objective lens using the cleaning solution and lens paper provided by your teacher. Remove your coverslip and put one drop of crystal violet stain under your coverslip and add more water if necessary. Put a new clean coverslip on your slide and focus your slide using the low-, medium-, and high-power objectives. Use the oil immersion objective as before and look at your slide. Can you see more bacterial cells after you added the crystal violet stain? How many different shapes of cells do you see?

12. Repeat steps 1 through 11 for Sample B.
13. Upon completion, consider the following:
    a. Were algae in any of your soil samples? From which place?

    b. Did you see fungal hyphae in any of your soil samples? From which sample?

    c. Did you see bacteria in any of your soil–water suspensions? In which sample(s)?

    d. In which of the two soil samples (A or B) were there more bacteria?

Questions for Thought

1. What would the presence of algae suggest about the amount of water the soil is holding?

2. What would fungi indicate about the type(s) of plants growing in the area?

3. What roles do fungi play in soils?

4. In terms of your original idea, how are plant cover and tilling related to the numbers and kinds of soil microorganisms?

Observing Legume Root Nodules

Plants do not grow well in soil with low levels of nitrogen. Many important agricultural crops such as wheat, corn, sorghum, cotton, tobacco, and oats require a lot of nitrogen. The practice of crop rotation, planting different crops in a field (or using the field as a pasture) from one growing season to the next, can greatly improve soil quality and reduce the amount of chemical fertilizer required for profitable yields. Growers who use crop rotation frequently use leguminous plants as an alternative crop to winter wheat, corn, and cotton. Nitrogen-fixing bacteria belonging to the genus ***Rhizobium*** are found in association with bumps (nodules) on the roots of peas, beans, clovers, and other leguminous plants.

Questions for Thought

1. In which of the soil samples (Sample A, under cultivation with leguminous plants, or Sample B from a pasture currently in use) would you expect the level of soil nitrogen to be greatest?

2. Explain how the amount of available soil nitrogen is related to the differences in cultivation practices among the sites.

Activity: Observing Legume Root Nodules

Materials Needed

- Legume root with visible nodules
- Dissecting needle
- Immersion oil
- Lens cleaning solution and lens paper
- Crystal violet stain

Procedure

1. Obtain a legume root with visible nodules from your teacher. Using a dissecting needle, put one or two nodules on a clean microscope slide.
2. Put several drops of water and one drop of crystal violet stain on the nodules.
3. Add a coverslip. Wrap the slide in a paper towel with the coverslip side up.
4. Using the eraser end of your pencil or the blunt end of your dissecting needle, crush the nodules until the coverslip is flat.
5. Using the low-, medium-, and high-power objectives, focus the slide.
6. Use the procedure for the oil immersion objective lens and look at the nodules at 1,000 × magnification. Draw what you see in the space provided. What shape are the *Rhizobium* bacteria (coccus, bacillus, or spirillum)?

Effects of Pea Inoculum on Legume Seedling Growth

Nitrogen-fixing bacteria help plants in the pea family (legumes) to grow in places where the soil is low in nitrogen. Many important agricultural crops are legumes, and commercial growers have found a way to increase the number of nitrogen-fixing bacteria in the roots of leguminous plants. Nitrogen-fixing bacteria are freeze-dried and added to a powder so that legume seeds can be easily coated with the beneficial bacteria. The mixture, called legume inoculum, can be purchased at most farm suppliers. When the coated seeds are planted, the soil moisture and warm soil temperatures activate the bacteria. As the seed germinates, the bacteria colonize the root of the seedling and begin to fix nitrogen. In this experiment, you compare the growth rates of two different groups of soybean seedlings. One group of seeds will be treated with pea inoculum before the seeds are planted. The other group will not be treated with pea inoculum prior to planting. You will measure your seedlings over three weeks and compare the average seedling height in the two groups of plants.

Questions for Thought

1. Which group of soybean seedlings do you think will grow faster? Why?

Activity: Effects of Pea Inoculum on Legume Seedling Growth

Materials Needed

- 20 soybeans
- Pea inoculum powder
- 20 peat pots or small plastic pots
- 20 "packing peanuts" or small stones
- Plant markers or marking pen
- Potting soil
- Nursery flat or tray
- Ruler

Procedure

1. You will work with two or three partners to complete this experiment. Count out 20 soybean seeds.
2. Separate the 20 seeds into two groups with 10 seeds each.

3. Get 20 peat pots or small plastic pots from your teacher. Place small stones or a plastic packing "peanut" in the bottom of each pot to help the soil drain better.
4. Fill each pot with planting soil and gently tap the soil down until the soil is ¼ inch from the top of the pot.
5. Mark 10 pots CONTROL and 10 pots BACTERIA. You can write on the plastic pots or use plant markers (wooden or plastic). If you use plant markers, make sure that you put the markers in the soil of each pot before you begin planting your seeds.
6. Using the eraser end of a pencil, make a hole in the soil of each pot about ½ inch deep.
7. Label your CONTROL pots 1 to 10.
8. Plant one untreated seed in each of the 10 CONTROL pots by putting a seed in the hole you made with the eraser. Cover the seed with soil. Place the 10 pots with untreated soybean seeds in a nursery flat full of water until the top of the soil is moist. Remove the pots from the water, and place them in a bright window or on a greenhouse bench.
9. Label your BACTERIA pots 1 to 10.
10. Put your 10 soybean seeds in a dish with the pea inoculum. Gently roll each of the seeds in the dusty powder until the surface of the seed is completely coated.
11. Drop one inoculated seed in the eraser hole in the soil of a BACTERIA pot. Cover the seed with soil. Repeat the process until all 10 inoculated seeds have been planted. Place the 10 pots with inoculated soybean seeds in a nursery flat full of water until the top of the soil is moist. Remove the pots from the water, and place them in a bright window or on a greenhouse bench.
12. When the first seedling breaks through the soil surface, begin measuring your seedlings and record the height of each seedling in centimeters (cm) in Table 4-3 (CONTROL) and Table 4-4 (BACTERIA). You will need to measure your plants every three days for three weeks.
13. Upon completion, consider the following:
    a. Which group of seeds germinated first?

    b. Which group of seedlings grew taller? (**HINT:** Compare the average height for the CONTROL group in Table 4-3 with the average height for the BACTERIA group in Table 4-4).

c. Did the results of your experiment turn out as you expected?

Name(s):					Date:			
Control	**Day 0**	**Day 3**	**Day 6**	**Day 9**	**Day 12**	**Day 15**	**Day 18**	**Day 21**
POT #1								
POT #2								
POT #3								
POT #4								
POT #5								
POT #6								
POT #7								
POT #8								
POT #9								
POT #10								
AVERAGE	[POT #1 + POT #2 + POT #3 + ... + POT #9 + POT #10]/10							

Table 4-3. Seedling height (cm) for CONTROL group soybean plants.

Name(s):					Date:			
Control	**Day 0**	**Day 3**	**Day 6**	**Day 9**	**Day 12**	**Day 15**	**Day 18**	**Day 21**
POT #1								
POT #2								
POT #3								
POT #4								
POT #5								
POT #6								
POT #7								
POT #8								
POT #9								
POT #10								
AVERAGE	[POT #1 + POT #2 + POT #3 + ... + POT #9 + POT #10]/10							

Table 4-4. Seedling height (cm) for BACTERIA group soybean plants.

Questions for Thought

1. Explain why you think the tallest group of seedlings grew better. How is the growth of the seedlings related to the inoculum?

Glossary

A horizon: the top mineral layer of the soil profile; contains the topsoil organic material mixed with the minerals originating from the parent rock

bacillus: (pl. *bacilli*) a rod-shaped bacterium

clay: the smallest particle size class in soil; less than 0.002 mm in diameter.

coccus: (pl. *cocci*) a round-shaped bacterium

detritus: loose pieces of material that come from the decomposition of organic matter

gravel: mineral pieces greater than 2.00 mm in diameter

humus: organic matter in soil that is decayed; humus has a high nitrogen (N) content

hyphae: (singular: *hyphus*) the multicellular threadlike filaments that make the body (mycelium) of a fungus

inorganic: a compound that lacks both the chemical elements carbon (C) and hydrogen (H); for example, sodium chloride (NaCl).

legume: the fruit or seed of a plant in the pea family. Leguminous plants include clover, peanuts, soybeans, peas, beans, alfalfa, lespedeza, and vetch. These plants aide in the return of organic nitrogen to the soil because their roots contain nitrogen-fixing bacteria.

loam: ideal soil texture for cultivation, consists of less than 52 percent sand, 28 to 50 percent silt, and 7 to 27 percent clay

mycorrhizae: a symbiotic relationship between a fungus and plant roots, usually a tree or shrub. Mycorrhizal relationships are more common in areas where the soil is poor in nutrients. The fungus allows the plant to take advantage of the reduced nutrient concentration in the soil.

organic: a compound that contains both the chemical elements carbon (C) and hydrogen (H) and comes from living matter; for example, methane (CH_4).

nitrogen fixation: the conversion of atmospheric nitrogen (N_2) to ammonia (NH_3) by certain bacteria associated with the roots of leguminous plants and cyanobacteria (blue-green algae) with heterocysts

ped: unit of soil structure; peds are described by their shape: aggregate, crumb, prism, block, plate, or granule.

pH: a measure of the relative concentration of hydronium (H^+) ions. The greater the H^+ concentration, the more acidic a substance is; the lower the concentration of H^+ ions, the more alkaline a substance is.

1		7		14
very acidic	acidic	neutral	alkaline	very alkaline

Rhizobium: a group of soil bacteria that penetrate the roots of plants in the pea family, forming nodules on the roots. Once inside in the root, these bacteria fix atmospheric nitrogen.

sand: mineral particles between 2.00 and 0.05 mm in diameter; may also be a textural class of soils that contain at least 85 percent sand and less than 15 percent clay particles.

silt: very small mineral particles in the soil in the USDA System, particles between 0.05 and 0.002 mm in diameter; also relates to the textural classification of soils—silty soil contains at least 80 percent silt and less than 20 percent clay particles.

slope: the angle of the land; calculated as the change of vertical distance with respect to change in horizontal distance; slope is expressed as a percentage:

$$\text{Slope} = \frac{\text{Vertical Distance}}{\text{Horizontal Distance}} \times 100 \text{ percent}$$

spirillum: (pl. *spirilla*) a corkscrew- or spiral-shaped bacterium

subsoil: the part of the soil that is below the cultivated layer; the portion of the soil profile in which a change occurs in the texture and appearance of the soil.

topsoil: surface and subsurface soils rich in organic material; fertile portion of the soil profile used for crop production.

EXERCISE

5

Parts of Plants

Student Objectives

After completing this lab activity, you should be able to:

- Define gymnosperms and angiosperms as types of seed plants, and describe the differences between them.
- Identify ways in which seed plants are important to humans.
- Identify the major parts of plants.
- Define the function of each of the major plant parts.
- Give examples of each of the major plant parts that are of agricultural importance.
- Describe the importance of photosynthesis in agricultural crops.
- Describe the development of plant root systems.
- Observe and describe the roles of xylem and phloem in the transport of water, minerals, and nutrients.
- Describe the role of the seed coat in germination.
- Observe and describe the parts of the seed.

Recommended Resources:

Chapter 4 in *Exploring Agriscience, 5th Edition.*

Introduction

Vascular plants have tissue that transports water, minerals, and food. **Seed plants** are vascular plants that reproduce by forming seeds. There are two main groups of seed plants: gymnosperms and angiosperms. A gymnosperm produces seeds that are not enclosed in a special structure such as a fruit. In many gymnosperms, like pine trees, seeds develop in the woody scales of a cone. An angiosperm produces seeds that are completely enclosed in a structure called a fruit. Angiosperms are flower-producing plants.

Seed plants are the most abundant plants on Earth, and people depend on them in many different ways. Seed plants provide cereal, bread, fruits, vegetables, and even perfumes. Lumber and paper products are made from seed plants. Fibers from cotton and flax are used to make clothes. Medicines, rubber, dyes, oils, and tobacco are also produced from seed plants.

Many parts of seed plants are useful to agriculturalists. **Roots** are usually found underground and anchor the plant in the soil and absorb water and minerals. Beets, carrots, sweet potatoes, and turnips are all plant roots. Some of the fastest-lighting kindling wood comes from the roots of plants.

Stems are the stalks, trunks, or branches of plants. They can be either vertical or horizontal, above or below ground. Many products are made from plant stems. Sugar cane, molasses, lumber, cloth, and potatoes are all either plant stems or are produced from plant stems. Plant stems are also used in feeding livestock. Corn stalks and grains such as alfalfa and sorghum are ground up and used as livestock feed.

Leaves are usually flat-shaped, green, and found at the end of stems. They are where most photosynthesis occurs. **Photosynthesis** is the process by which the plant produces food. Because leaves make food for the plant, they tend to be very nutritious for people and other animals. Lettuce, spinach, and even onions are plant leaves. Many livestock animals graze primarily on the small leaves of grass plants. Leaves are also useful to humans in other ways. Tobacco products are produced from the leaves of the tobacco plant. Pound for pound, mulched tree leaves have twice the minerals of manure. Mulched leaves can also dramatically improve the structure of soils. The mulch helps aerate heavy clay soils, soaks up water, and prevents evaporation in sandy soils.

Seeds are reproductive structures produced by flowers. Each seed contains an embryo, stored food, and a seed coat. Upon formation, some seeds are able to germinate immediately, while others require some extreme conditions such as a cold season, exposure to fire, or even exposure to animal digestive juices before they are able to germinate.

Activity: Plant Roots and How They Develop

Materials Needed

- Plastic bottle
- Scissors
- 1 quart of dried soil
- Water
- 12 × 24-inch window glass
- 11 × 23-inch sheet of cellophane or thin plastic
- Corn, cotton, soybean, or other crop seed
- Black plastic or paper

- Bricks or blocks
- Mirror
- Plastic bag and rubber band
- Wax pencils in several colors
- Ruler

Procedure

1. Cut the top from a plastic bottle just below the shoulder (see Figure 5-1). Fashion three legs ¼ inch high on its lower rim.
2. Pulverize one quart of soil. Slowly add water until the soil can be formed into a ball (not muddy, but moist enough to hold).
3. Crumble and spread about two cups of this moist soil evenly on a 12 × 24-inch window glass, leaving 1 inch around the edge free of soil.
4. Make an X-shaped cut in the center of an 11 × 23-inch sheet of cellophane or thin plastic (see Figure 5-2a).
5. Fit this hole about the neck of the bottle and center it on the soil-covered glass. Tape the plastic to the neck of the bottle, as shown in Figure 5-2b. Carefully spread the plastic sheet over the soil and fasten the edges to the glass with tape.
6. Fill the bottle top with moistened soil and insert one seed. Use corn, cotton, soybean, or another crop seed.
7. Press the seed 1 inch into the moistened soil and cover. Keep moistened.
8. Place in a well-lighted spot but not in direct sunlight or near a source of heat.
9. To prevent mold growth, cover soil area around the hole with black plastic or paper.
10. Place window glass on blocks or bricks, as illustrated in Figure 5-3, with a mirror below to observe the growth of roots.
11. Check daily for signs of roots. They usually appear in four or five days at a temperature around 80°F. Record your daily observations.
12. As the plant begins to grow, place a plastic bag over the top of the plant and seal it at the neck of the bottle with a rubber band.
13. Once a week, draw the outline of the root system on the bottom side of the glass (see Figure 5-4). Use a different colored wax pencil each week.

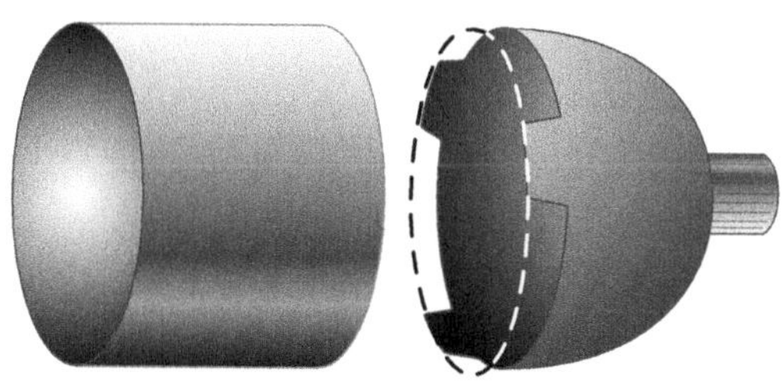

Figure 5-1. Cutting the soda bottle.

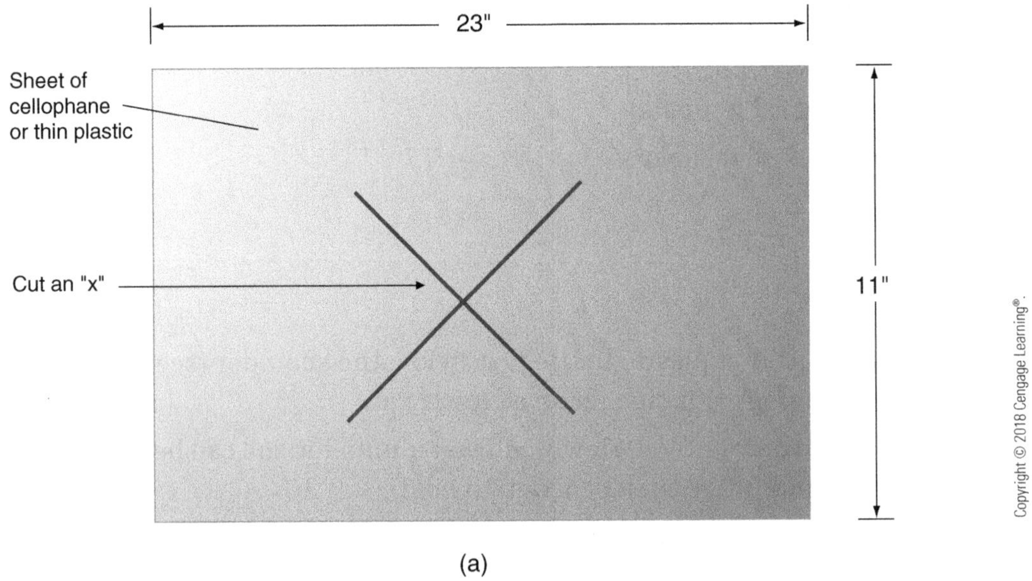

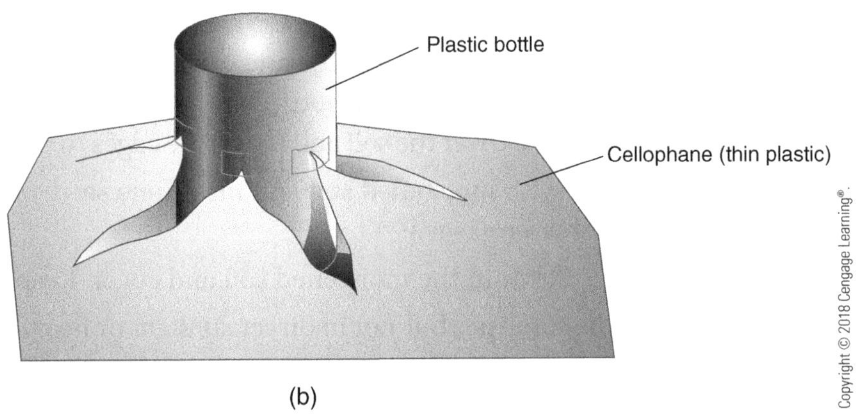

Figure 5-2a and b. Attaching the plastic to the bottle.

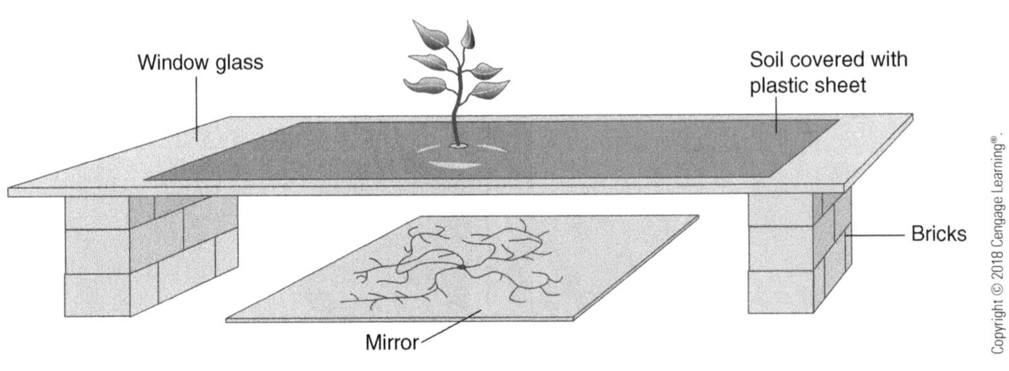

Figure 5-3. Setting up the root observation.

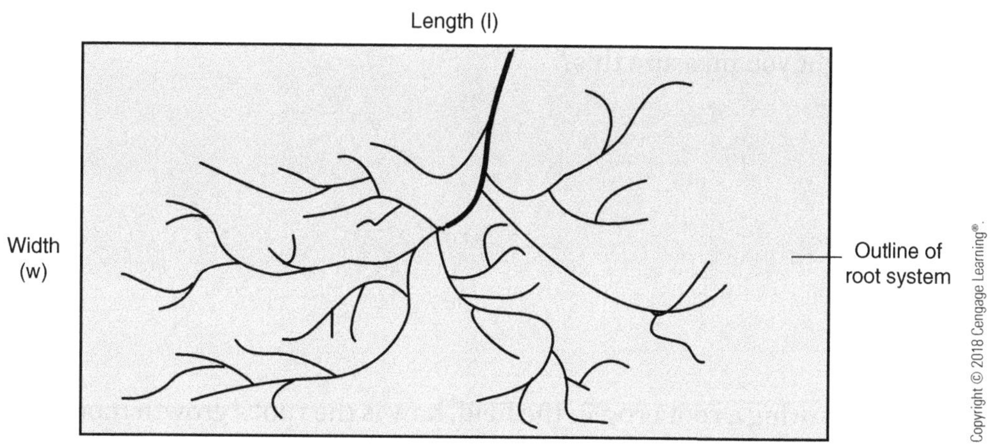

Figure 5-4. Measuring the area of root growth.

14. Using a ruler, draw the smallest possible fitting box around the root outline. Using the formula length × width = area, calculate the approximate area of the root system.
15. At the end of five weeks, compare the root drawings. Calculate the percentage of root area increase for each week and record your findings in Table 5-1.

Planting Date:	Date Seedling Emerged:	Date of First Roots:
	Approximate Area of Roots	% Increase Per Week
Week 1		
Week 2		
Week 3		
Week 4		
Week 5		

Table 5-1. Root observation.

Questions for Thought

1. What is the reason for covering the top part of the plant with plastic?

2. How does the root growth compare to the aboveground growth of the plant? How might you measure this?

3. When growing a root crop in the field, how is the root's growth monitored?

Plant Stems and Water Transport

Tissues called xylem and phloem are tiny tubelike structures that act as a transport system for water and nutrients in plants. **Phloem** transports sap, which is composed of sugars manufactured in the leaves, all through the plant for growth and storage. **Xylem** carries the water and minerals taken up by the roots to the plant's stems and leaves. **Osmosis** is the movement of the water from the soil to the xylem inside the root. Water will move from an area where it is plentiful into an area where it is less plentiful. In this way, as water moves from the roots into the stem, the roots become dry and water can move into them easily. In tiny tubes like xylem, capillary action causes water to rise against the pull of gravity. The rising water can be attributed to the attraction between the water molecules and the sides of the tube.

Activity: Water Transport

Materials Needed

- Small glass or rubber tube
- Rubber tubing (in a size that will fit over the above tube)
- Shoot from a plant such as bean, sunflower, begonia, dahlia, or corn
- Shears or a knife
- Salt
- A one-gallon container
- Marking pen

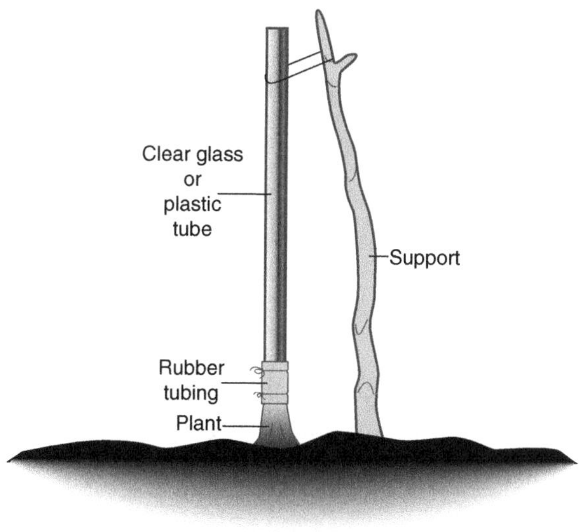

Figure 5-5. Attaching the tube and rubber tubing to the plant stump.

Procedure

1. Get a small glass or clear plastic tube and a short piece of rubber tubing that will fit over it.
2. Select a shrub or tree shoot that matches the size of the rubber tube. Plants like beans, sunflower, begonia, dahlia, and corn will work.
3. Cut off the plant near the ground. If water (sap) does not come from the cut, choose another plant.
4. Attach tube and rubber tubing to the plant stump, as shown in Figure 5-5.
5. Put enough water into the tube to bring the level above the rubber tubing.
6. A drop of oil on the water in the tube will help you note the level of the liquid.
7. Place a mark on the tube at the oil level. Record the time.
8. Record the changes in the oil level hourly until it stops rising.
9. Add table salt to one gallon of water until no more will dissolve when stirred, and pour this solution slowly around the plant.
10. Mark the tube at the beginning of the salt water experiment and record in Table 5-2 the water level hourly until it stops falling.
11. Calculate the average speed in inches per minute that the water traveled up and down the tube.

	Plain Water	Salt Water
Hour 1 Water Level		
Hour 2 Water Level		
Hour 3 Water Level		

Table 5-2. Water levels.

Questions for Thought

1. Why did the water move up the tube when the plant stem was first cut?

2. Why did the water move down the tube when salt water was added to the soil? Under what circumstances might this happen in nature? What consequences would it have for the plant?

Leaves Evaporate Water from Their Surfaces

Stomata are usually found on the lower surfaces of leaves. They are tiny openings that allow gases to move in and out of the leaf. Carbon dioxide passes into the leaf, and water vapor and oxygen pass out of the leaf through these openings. **Transpiration** is the process by which leaves release water vapor into the surrounding atmosphere.

Activity: Transpiration

Materials Needed

- A fresh green leaf
- Two pieces of cardboard
- Four cups or glasses

Procedure

1. Collect a fresh green leaf about 1½ to 2½ inches long.
2. Cut two pieces of cardboard to cover water glasses, as shown in Figure 5-6.
3. Punch a small hole in the center of each piece of cardboard.
4. Fill the two glasses with water.

5. Place the cardboards over the full glasses, insert the leaf stem through the hole in one, and into the water below.
6. Put empty glasses upside down over the cardboards.
7. Set in a warm area and observe.
8. Compare the two top glasses after 45 minutes. What is the difference?

Questions for Thought

1. Why do you think the water formed in the glass with the leaf? Where did it come from and how did it get there?

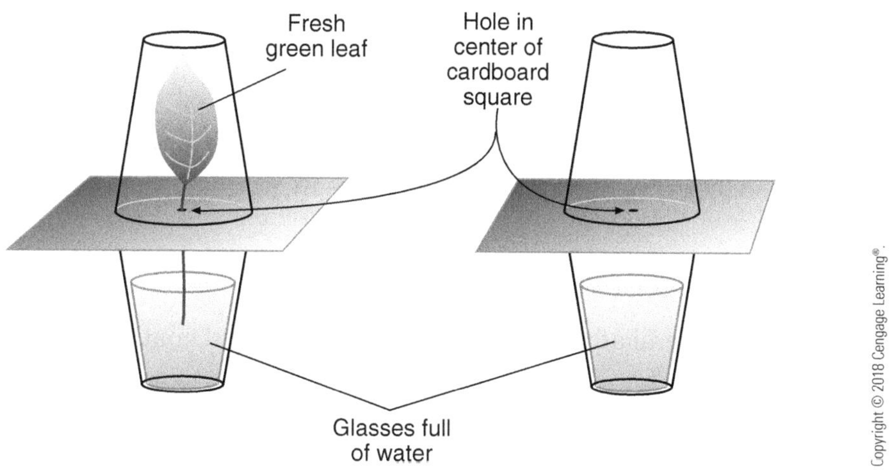

Figure 5-6. Setting up transpiration experiment.

2. How do you know that the water did not form from evaporation alone?

Seed Coats

A tiny, multicellular developing young plant is called an **embryo**. In seed plants, the embryo develops inside a seed. The seed consists of an embryo, stored food, and a seed coat. The **seed coat** is a tough outer covering that protects the seed.

Activity: Seed Coats

Materials Needed

- Two teaspoons of a hard seed (examples: crown vetch, crimson clover, mimosa, or morning glory)
- Boiling water
- A jar with sandpaper strips
- A needle
- Four pots with potting soil
- Plant markers

Procedure

1. Select a hard seed such as crown vetch, crimson clover, mimosa, or morning glory.
2. Divide the seeds into four groups of about ½ teaspoon each and number them 1 to 4. Treat group #1 by placing in boiling water for 3 minutes; treat group #2 by shaking in a jar with sandpaper strips; treat group #3 by piercing each seed coat with a needle; and leave #4 untreated.
3. Plant the groups of seeds in pots labeled with the type of treatment they received. Observe and record differences in germination.

Questions for Thought

1. What are several advantages of having a hard seed coat?

2. Can you think of any disadvantages of having a hard seed coat?

3. Could seed coats pose problems for planting and growing crops successfully?

4. How might these problems with seed coats be addressed?

Parts of a Seed

Some seeds can be easily split in half, which allows us to investigate their contents. When you open the seed, usually you can see a miniature plant (embryo) and the seed coat. (see Figure 5-7).

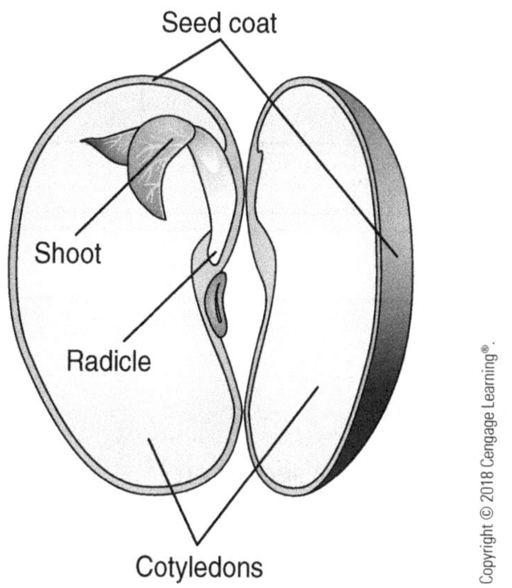

Figure 5-7. Parts of a Seed

Activity: Seed Parts

Materials Needed

- Seeds for corn, lima beans, sunflowers, and peanuts
- A container of water for soaking the seeds
- A scalpel

Procedure

1. Soak corn, lima beans, peas, peanuts, and sunflower seeds in water for 24 hours.
2. Gently use your fingernail to peel the softened seed coat from a soaked lima bean. Split the seed in half and identify the parts labeled: embryo with seed leaves, seed coat, stored food.

3. Remove the seed coats from the peanut, corn, sunflower seed, and pea. Split each seed in half. If the seeds do not open easily, use a scalpel to carefully slice the seed open lengthwise. Record your observations in Table 5-3.

CAUTION:
Be very careful when using the scalpel to avoid cutting yourself.

Type of Seed	Seed Coat (Hard or Soft)	Easily Opened?	Number of Leaves (1 or 2)	Future Root Present?
Lima Bean				
Peanut				
Sunflower Seed				
Corn				
Pea				

Table 5-3. Seed observation.

Questions for Thought

1. What is the advantage of the seed coat?

2. Why does the seed have a supply of stored food?

3. In what ways is a seed similar to a chicken egg? In what ways is it different?

4. Were any of the seeds lacking future leaves or roots? What would happen to the plant if these structures were lacking?

Glossary

embryo: a plant in the early stages of development

leaf: growth off a stem, usually flat and green. Primarily concerned with food production.

osmosis: the movement of the water from the soil to the xylem inside the root.

phloem: plant tissue responsible for the transport of food produced in the leaves to other parts of the plant

photosynthesis: process by which green plants use sunlight, water, and carbon dioxide to make food and oxygen

root: the lower part of the plant that bears no leaves or reproductive organs. It develops mostly underground and absorbs water and minerals and anchors the plant.

seed: plant embryo and stored food surrounded by a seed coat

seed coat: a tough outer covering that protects the seed

seed plant: a vascular plant that reproduces by forming seeds

stem: stalk, branch, or trunk of a plant

stomate (plural: *stomata*)**:** an opening into an internal air cavity below the skin of the leaf

transpiration: process by which water vapor is released from the leaves of a plant

vascular plants: a plant that has special tissue through which water and nutrients are transported

xylem: plant tissue responsible for transporting water and minerals from the roots to other parts of the plant

EXERCISE 6

Biological Control of Insect Pests

Student Objectives

After completing this lab activity, you should be able to:

- Explain why certain insects are described as pests.
- Construct a leaf press and use it to preserve leaves.
- Distinguish between the three main types of biological control of insects.
- Prepare an insect collection.

Recommended Resources:

Chapter 5 in *Exploring Agriscience, 5th Edition.*

Introduction

There are three major components of biological control—**predators**, **parasites**, and **pathogens**. The difference between natural control and biological control is that biological control is in some way influenced by humans.

The list of animals that are predators of insects is long. It includes fish, frogs, lizards, birds, bats, badgers, skunks, and other animals. In many instances, humans use insects in controlling pests. For example, praying mantises and ladybugs are living biological control agents.

Like humans, **insects** harbor diseases. Insects can be affected by specific protozoa, fungi, viruses, and bacteria. **Viruses** and **bacteria** are the pathogens most commonly used in biological control efforts.

Insects make up about three-fourths of all animal species. The number of insects is greater than any other type of animal. Therefore, many species may be found, caught, and examined very easily.

Insects have bodies supported by an **exoskeleton.** This is a skeleton on the outside of their bodies and acts as a

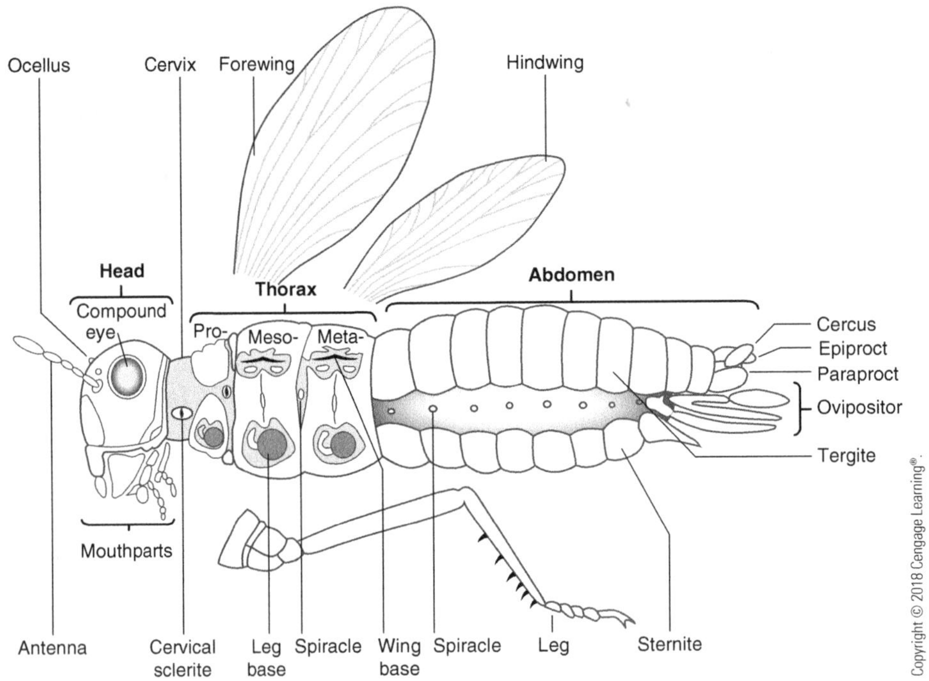

Figure 6-1. The generalized body plan of a winged insect.

protective covering. Insects have three body parts. The head holds the eyes, mouth parts, and antenna. The thorax is the middle part where the legs and wings are attached. The abdomen is behind the thorax and contains the organs of digestion and reproduction. Insects have three pairs of legs and antennae on the front of the head. These serve primarily as organs of touch (see Figure 6-1).

Activity: Insect Collection

Materials Needed

- Net with fine mesh and handle
- Cotton swab
- Jar or similar container with lid
- Rubbing alcohol
- Straight pins
- Shoe box or foam board for mounting specimens

Procedure

1. Using the net, catch as many different insects as possible in the time allotted.
2. Dip the cotton swab in alcohol and place it in the bottom of the jar.
3. Place the insects in the jar and quickly seal the lid. This will kill the insects so that they can be handled more easily.

NOTE: Insects many be found in different places from spring to fall. Look for insects around blooming plants, leafy plants, water, and fallen trees.

4. After you have collected the insects, bring them into class. Identify 10 different specimens from the collection.
5. In the space provided, determine whether each specimen is an insect or an arachnid. (**HINT:** Insects have six legs, and arachnids have eight legs. Insects have three distinct body regions in contrast to the two body regions found in arachnids.)

6. After you have identified and named the specimens, use pins to spear each specimen through the thorax (Figure 6-1). Neatly mount the specimens in a shoe box or on foam board and label them.
7. Determine the ratio of insects to arachnids.

Insect Damage

Many insects can damage trees and other plants by eating their leaves for food. This reduces a plant's ability to photosynthesize, so the plant may suffer loss of growth or even die. Insects that destroy other insects are divided into two groups known as predators and parasites. Predators catch and eat their prey, usually eating them in one meal. Parasites live in or on the bodies of living organisms called **hosts** from which they get their food during at least one stage of their lives. Predators are usually very active and have long life cycles. Parasites are usually sluggish and have short life cycles.

Activity: Investigate Insect Damage

Materials Needed

- Two 14 × 18-inch boards
- String or strap to hold the press closed
- 10 sheets of cardboard
- Newspapers
- Index cards
- A photo album with peel-back clear pages

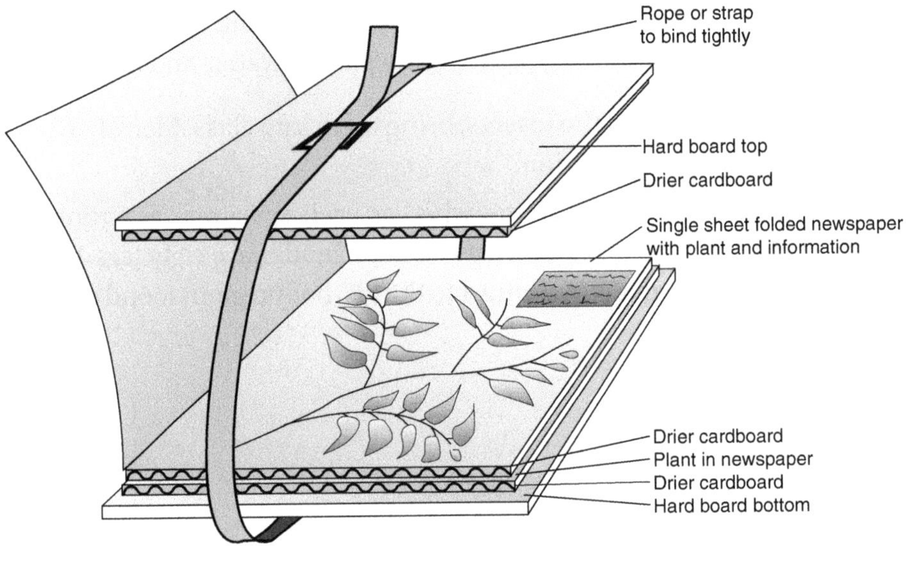

Figure 6-2. A plant press.

Procedure

1. Visit any nearby woods or forest and examine the trees there. The most obvious damage may be on the leaves. Collect 10 damaged leaves in the area where the insects were collected.
2. Construct a leaf press as outlined in Figure 6-2. Materials will include two 14 × 18-inch boards, a string or strap to hold everything closed, 10 sheets of cardboard, many newspapers, and a photo album with peel-back clear pages.
3. Record where each of the leaves was found, date, tree species, and your name on an index card and place it with the specimen in the leaf press.
4. Draw one of the damaged leaves on graph paper and determine the percentage of leaf eaten by the insect.

Questions for Thought

1. Did you find the damage to the leaves? Try to determine what type of insect pest caused the damage. How do you know?

2. Examine the bark. List observations about the bark in the space provided. If the bark is loose or has fallen off, look at its inside and at the layer of wood underneath the bark for signs of other damage.

3. Look at the top shoot of a small tree. Record in the space provided whether the shoot is growing straight up or not.

4. If it is bent or twisted, this damage could be caused by stem feeders. From your observations, describe how you think stem feeders damage the plant.

5. Examine a variety of trees. Look at different species, and at both young and old trees. Do you see a pattern in which trees were eaten by the insects? Why do you think that this is so?

6. Patches of dead trees in a forest sometimes indicate insect damage. Do you see such areas? If so, do you feel this area was damaged by insects? How can you tell?

7. Could the trees have been killed by something else? If you think so, what could it be?

Glossary

bacteria: one-celled microorganisms that have no distinct nucleus; bacteria have a cell wall and a cell membrane.

exoskeleton: any hard, external, secreted supporting structure

host: organisms from which a parasite takes its nourishment

insects: class of small arthropods with segmented bodies and three pairs of legs

parasite: any animal that lives in, or on, or at the expense of another

pathogen: disease-causing organism

predator: an organism that lives by feeding on others

viruses: nucleic acids containing prokaryotes which replicate in other living cells

EXERCISE

7

Cultivating Crops—Composting and Animal Pests

Student Objectives

After completing this lab activity, you should be able to:

- Explain the purpose of composted material and its relationship to soil management.
- Construct a compost column.
- Contrast the effects of certain factors on composted material.
- Assess the amount of damage on land due to various animal pests.

Recommended Resources:

Read Chapters 3, 5, and 18 in *Exploring Agriscience, 5th Edition.*

Introduction

Compost is the organic end product of the decaying process. When dead material, either plant or animal, **decomposes**, it returns nutrients to the soil. A compost heap is a collection of organic materials such as leaves, grass, and uncooked foods that will decompose over time to create rich, **fertile** soil.

Compost material contains many kinds of organic materials. Things that make for efficient compost materials include grass clippings, leaves, coffee grounds, tea bags, cotton, twigs, and weeds. Some things not to include are pet manure, cooked foods, and meat.

The compost pile decomposes through the action of **microorganisms**. When dirt is shoveled on the pile, the microorganisms will immediately begin decomposing the material in the pile. Turning the pile once weekly keeps it active and well aerated.

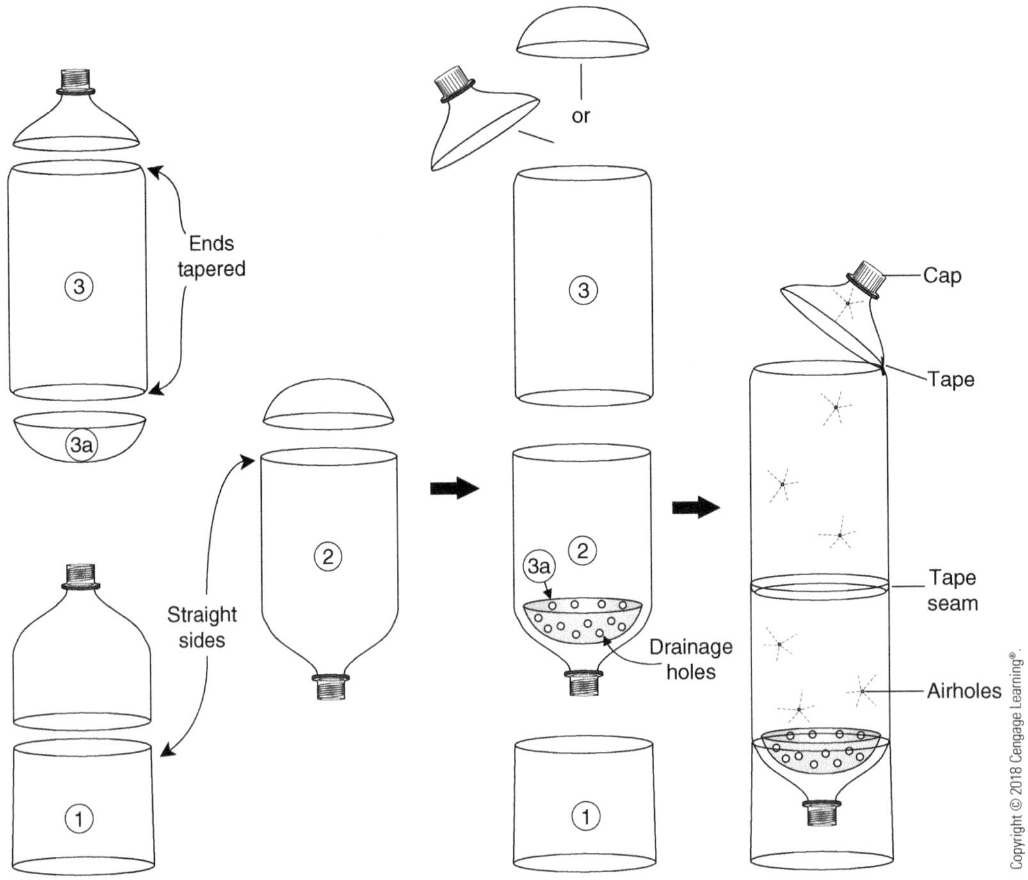

Figure 7-1. Compost column construction.

Many agricultural pests such as **fungi** and **nematodes** can be controlled by soil management. Using completely composted plant material reduces the chances of spreading fungal diseases from the mulch to new plants. The amounts of moisture, air, temperature, light, sources of bacteria, and fungi are examples of factors that affect the composting process. The presence of oxygen is one of the most important factors. Composting allows air and moisture to speed the natural process of decomposition.

Activity: Making a Compost Column

By making a compost column, as shown in Figure 7-1, you can see what happens in this natural process.

Materials Needed

- Three 2-liter bottles
- Scissors
- Marking pen
- Clear tape
- Mesh fabric

- Rubber bands
- Organic material: leaves, newspapers, grass clippings, etc.
- Optional: a tray of seedlings to compare growth

Procedure

1. Remove the labels from all three bottles. (Heating the label with warm water may help.)
2. Draw cutting lines around the bottles. Cut and assemble, as shown in Figure 7-1.
3. Most columns will require air holes for ventilation. A piece of mesh fabric over the lower end allows for drainage.
4. Add ingredients for composting through the top of the column.
5. Your class will be divided into five groups. Each group will alter one of the following conditions: temperature, light, type of compost material, shaking (to ensure a high level of oxygen), and moisture.
6. Each day, take a measurement of the height of the compost inside the column and record this information along with visual observations of the changes in the column. Also record the temperature of the column. This compost can be saved for Exercise 13.

   NOTE: Compost columns can be used to generate a liquid **fertilizer**. Test the effects of this liquid fertilizer by adding it to half of a tray of seedlings while the other half receives only soil and water. Compare the growth of these seedlings with the growth of the seedlings grown only in soil and water. Record your observations of what happens to the seedlings for two weeks.

Questions for Thought

1. Describe how the alteration of each of the different conditions (temperature, light, type of compost material, shaking and moisture) affected the process of decomposition.

2. Describe the difference in the seedlings that received the liquid fertilizer, and those that did not.

Identifying Animal Pests

Although observing wildlife in their natural habitat can be fun and enjoyable, there are certain times when certain species are not welcome or beneficial. In many areas animals can act as a nuisance. A good example of this is the white-tailed deer, which may feast on crops, gardens, and landscapes, and often over-populate a region. Signs of damage, including chewed leaves, animal prints, and other such evidence, can be the key to identifying animal pests.

Activity: Evidence of Animal Pests

As a class group, observe a wooded area, a plot of grazing land, and a garden. In journal, record evidence of animal pests to the area.

Materials Needed

- Notebook/Journal
- Pen/Pencil

Questions for Thought

1. If there is evidence of animals such as deer, does it seem to affect the growth in the area?

2. If so, is there too much damage to produce an effective yield?

3. What could be done to control the animals if too much damage was present?

Glossary

aerated: mixed with air

compost: a mixture of vegetable matter used to fertilize land

decompose: to break up into component parts

fertile: highly productive

fertilizer: a preparation containing elements needed for plant growth

fungi: parasitic or saprophytic organisms without chlorophyll; fungal cells have cell walls and cell membranes

microorganism: any organism that is not visible with the unaided eye

nematodes: microscopic unsegmented worms

EXERCISE

8

Controlling Plant Flowering

Student Objectives

After completing this lab activity, you should be able to:

- Describe photoperiodism based on laboratory experiences.
- Describe which part of the plant senses daylight based on laboratory experiences.

Recommended Resources:

Read Chapters 4 and 6 in *Exploring Agriscience, 5th Edition.*

Introduction

Plants and animals adapt to the places where they grow and live in many ways. Many plants and animals respond to seasonal changes in the weather. During autumn in the northern hemisphere, days grow shorter and nights grow longer. In spring, the days grow longer and the nights grow shorter. Plants respond to these changes in daylength. Changes in the **daylength** tell them when to rest and when to flower. In autumn, the leaves of some trees turn bright colors and fall off while the buds for next year's growth become inactive **dormant**. In spring, many different species of plants bloom and set seeds or fruit.

In 1920, W. W. Garner and H. A. Allard discovered that changes in the daylength, or **photoperiod**, affect the number of days it takes before the plant flowers. The number of hours of light and dark in one 24-hour period is called a photoperiod. Plants can be divided into three groups depending on how many hours of light and dark they need in order to flower. The three groups of plants are called **short-day plants** (SDP), **long-day plants** (LDP), and **day-neutral plants** (DNP) (see Table 8-1). Short-day plants flower when the daylength is shorter than a certain number of hours called the

critical daylength. See Table 8-2 for critical daylengths of some short-day plants. Long-day plants flower when the daylength is longer than the critical daylength. Day-neutral plants are not affected by daylength. Scientists also discovered

SHORT-DAY PLANTS	
Chrysanthemum	*Chrysanthemum morifolium*
Cocklebur	*Xanthium strumarium*
Biloxi soybean	*Glycine max* cv. Biloxi
Japanese morning glory	*Pharbitis nil* strain
Violet Maryland Mammoth tobacco	*Nicotiana tabacum* cv. Maryland Mammoth
Cotton	*Gossypium hirsutum*
Strawberry	*Fragaria spp.*
LONG-DAY PLANTS	
Sugar beet	*Beta vulgaris*
Winter barley	*Hordeum vulgare*
Spinach	*Spinacia oleracea*
Oat	*Avena sativa*
Rye-grass	*Lolium temulentum*
Radish	*Raphanus sativus*
Clover	*Trifolium spp.*
Mustard	*Brassica spp.*
Lettuce	*Lactuca sativa* cv. Black-seeded Simpson
Petunia	*Petunia hybrid*
DAY-NEUTRAL PLANTS	
Cucumber	*Cucumis sativus*
Corn	*Zea mays*
Tomato	*Lycopersicon sp.*

Table 8-1. Flowering behavior of some selected plants.

Day Plants	
Maryland Mammoth tobacco	12 hours
Biloxi soybean	14 hours
Japanese morning glory	5 hours
Cocklebur	15.5 hours
Chrysanthemum	13 hours

Table 8-2. Critical daylength of selected short-day plants.

that darkness is the most important part of the photoperiod. They found that darkness was so important that even a single flash of light during the dark period could keep a plant from flowering.

Activity: Effects of Short Daylength

Materials Needed

- Three flats or 2" × 2" pots
- Vermiculite or sand
- Short-day plants or seeds (see Table 8-2)
- Automatic timer
- Fluorescent lights

Procedure

1. Fill three flats with vermiculite or sand. Individual pots (2" × 2") can also be used.
2. Plant the flats with a suitable short-day plant (see Table 8-1). Biloxi soybeans work well (use pea inoculant when planting legume seeds).
3. For the next six weeks, provide a 20-hour photoperiod (4 A.M. to 12 P.M.) and a 4-hour dark period for the plants. Use an automatic timer and fluorescent lights to ensure accurate timing.
4. Adjust the lighting schedule so that the plants are exposed to less than the critical daylength. See Table 8-2 for a list of appropriate daylengths.
5. Leave one flat of plants under 20 hours light/4 hours dark as a control. Expose one flat of plants to the proper short-day lighting schedule for 10 days to two weeks. Expose another flat of plants to the same short-day lighting schedule for 10 days to two weeks, but expose the plants to 15 to 60 minutes of light in the middle of each dark period. This is fairly simple to do if the first flat is

Light Regime	# plants in group	# plants flowering	# flowers on flowering plants	% of flowering plants out of total
Control				
Short-Day (SD) Light				
Short-Day (SD) Light, Interrupted Night				

Table 8-3 Measurements of flowering.

either covered with a box or moved to a dark cabinet each night. Adjust the automatic timer so that it turns on each night for the desired period of time.

6. Return the plants to the regular lighting schedule for four weeks (20 hours light/4 hours dark).
7. Record your observations in Table 8-3.
8. Upon completion, consider the following:
   a. What percentage of the plants in the control group were flowering?

   b. What percentage of the plants exposed to short daylight were flowering?

   c. What percentage of the plants in the interrupted night group were flowering?

Questions for Thought

1. Why do you think there is a difference between the two groups?

2. Of the interrupted night group, why do you think they were or were not flowering?

What Part of the Plant Senses Daylight?

Animals have eyes that allow them to see their surroundings, but how do plants sense their surroundings? Although plants do not actually have eyes, they can sense whether it is light or dark and how long it has been light or dark. After the discovery of **photoperiodism** by Garner and Allard in 1920, scientists began asking many questions. One of the most interesting questions was, "What part of the plant knows how long the days and nights are?" A Russian scientist named M. Kh. Chailachjan conducted a series of experiments to determine whether the shoots or the leaves of a plant sense that the daylength is correct for flowering. See Figure 8-1 to better understand Chailachjan's experiments.

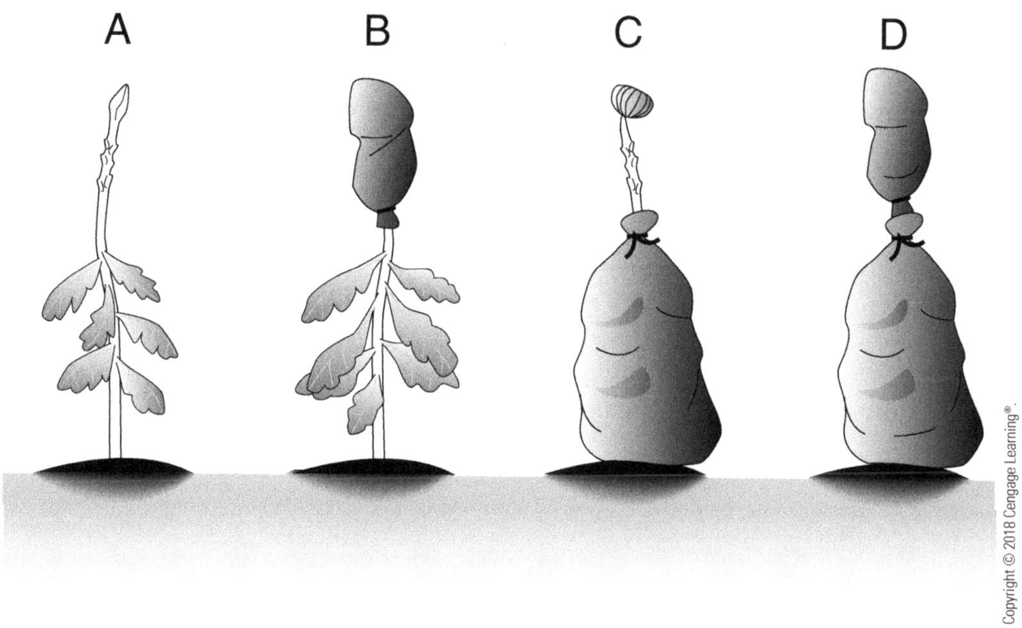

Figure 8-1. Chailachjan's lighting experiment (bags equal short-day light).

Activity: Plant Parts That Sense Daylight

Materials Needed

- 24 short-day plants grown for six weeks under 20 hours light/4 hours dark
- Dark plastic bags or paper bags
- Ties or tape for holding bags on stems
- Automatic timer
- Fluorescent lights

Procedure

1. Leave six plants as controls (as demonstrated in Figure 8-1a) (20 hours light/4 hours dark).
2. Cover the tops only (as demonstrated in Figure 8-1b) of six plants with either a dark plastic bag or a paper bag tied or taped closely to the stem (without damaging it).
3. Cover just the leaves (as demonstrated in Figure 8-1c) (leave the top of the plant exposed) of six plants with a dark plastic bag or paper bag or cardboard box cut in half vertically and taped around the plant.
4. Cover both the tops and leaves of six entire plants (exposed to short-day light, as demonstrated in Figure 8-1d).
5. Place all experimental group plants under appropriate SD lighting (as described in Table 8-2) for two weeks.
6. Return the experimental groups to regular lighting for four weeks or until flowering begins.
7. Record your observations in Table 8-4.
8. Upon completion, consider the following:
   a. How many plants in the covered tops group were flowering?

   b. How many plants in the covered leaves group were flowering?

Exposure Regime	# plants in group	# plants flowering	# flowers on flowering plants	% of flowering plants out of total
Control plants (regular light)				
Plants with covered tops (SD light)				
Plants with covered leaves (SD light)				
Control plants (SD light)				

Table 8-4 Determining which part of the plant senses daylength.

Questions for Thought

1. Why do you think there is a difference between these two groups?

2. Can you tell which part of the plant senses the daylength? Please explain.

Glossary

critical daylength: period of time during which a plant must exposed to light before flowering will occur

daylength: period of time during which a plant is exposed to light

day-neutral plant: plants that flower regardless of the daylength

dormant: a period of time during which growth ceases; growth resumes only when temperature and daylength requirements have been met

long-day plant: plant that flowers in response to daylengths longer than the critical daylength as determined by the individual species

photoperiod: number of hours of darkness and light in one 24-hour period

photoperiodism: the study of the behavioral responses of plants and animals to day- and night-length changes

short-day plant: plant that flowers in response to daylengths shorter than the critical daylength as determined by the individual species

References

Bernier, G., Kinet, J-M., and Sachs, R.M. (1981). *The physiology of flowering*, Vol. 1. Boca Raton, FL: CRC Press, 21–82.

Kendrik, R.E. (1981). How plants make light work of growth. *Journal of Biological Education*. 15(2): 85–89.

Moore, C.T. *Research experiences in plant physiology*. New York: Springer-Verlag, 345–59.

Noggle, G.R., and Fritz, G.J. (1983). *Introductory plant physiology*. Englewood Cliffs, NJ: Prentice Hall, 570–609.

Wareing, P.F., and Phillips, I.D.J. *The control of growth and differentiation in plants*. Elmsford, NY: Pergamon Press, 199–252.

EXERCISE 9

Asexual Propagation of Landscape Plants

Student Objectives

After completing this lab activity, you should be able to:

- Determine the effects of auxin on bulblet formation through laboratory investigations.
- Determine the effects of auxin on plant cuttings through laboratory investigations.
- Determine the best method for layering plants through experimental investigations.

Recommended Resources:

Read Chapters 4, 6, 7, and 9 in *Exploring Agriscience, 5th Edition.*

Introduction

Plant nurseries need to produce large numbers of healthy plants as quickly as possible in order to make the highest profits. Shrubs and trees take a long time to grow to a size that is useful for landscaping if they are grown from seed, so plant nurseries grow the plants through **asexual reproduction**. Asexual reproduction is the growing of plants through some means other than seeds. The five major methods of asexual propagation used to multiply plants are

1. division
2. cuttings
3. layering
4. **grafting**
5. **micropropagation**

Division

Plants can be multiplied by splitting a closely growing clump of plants or separating new plants from a "mother" plant. Almost all bulbs grown for sale are produced by division. Hyacinths have a unique ability to form new **bulblets** after wounding. Growers take advantage of growth substances that can be applied to plants to make them form roots or shoots more quickly.

Activity: Division Method

Materials Needed

- Six hyacinth bulbs
- Sharp knife
- Spoon or melon baller
- A tray
- Sand
- Auxin and lanolin paste

Procedure

1. Set two bulbs aside as controls in order to compare the formation of bulblets in the next two treatments to untreated or uncut bulbs.

   CAUTION:
   Be very careful when using a sharp knife to avoid cutting yourself.

2. To encourage the growth of new bulblets, notch the bottom of two hyacinth bulbs in a pie formation with a sharp knife (see Figure 9-1). Make your cuts about ¼ inch deep. Your instructor may choose to do this for you.
3. Another way to encourage the growth of new bulblets is to scoop out a portion of the base of the bulb. This is where new leaves originate. Scoop out the base of two hyacinth bulbs with a melon baller or spoon (see Figure 9-2).
4. Place all bulbs bottom-side-up in a moist tray of sand at room temperature. Treat one control, one notched, and one scooped bulb with a plant growth substance (auxin) by applying a thin film of lanolin paste to the cut areas.
5. Keep a progress record in Table 9-1.

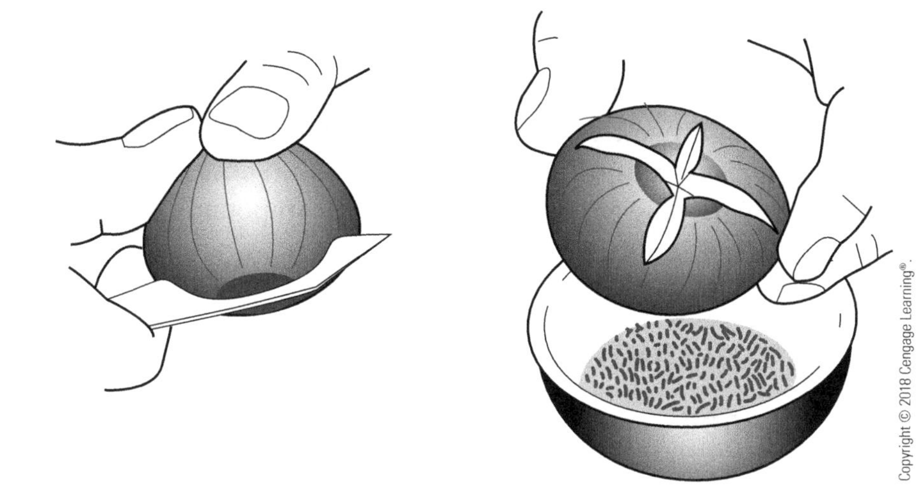

Figure 9-1. Notching a bulb.

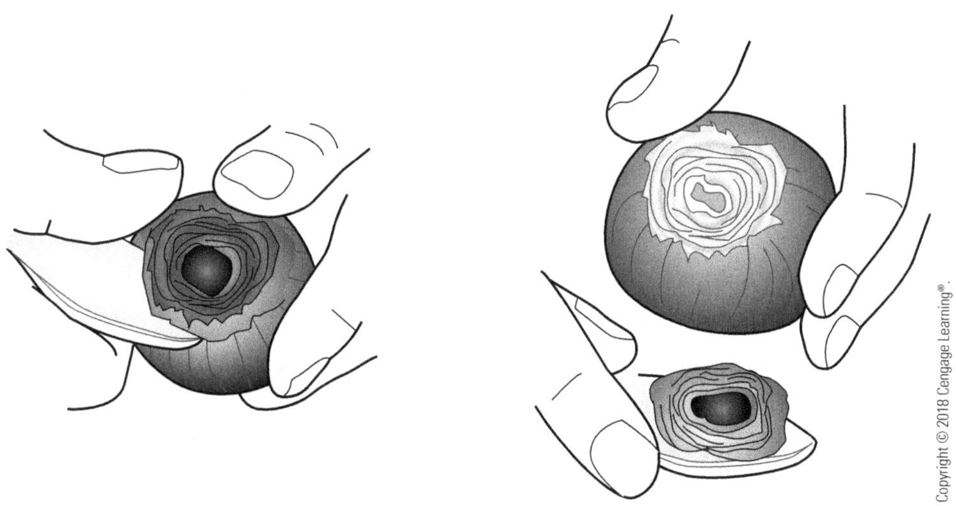

Figure 9-2. Scooping a bulb.

Bulbs	Water Only	Growth Substance
Control		
Notched		
Scooped		

Copyright © 2018 Cengage Learning®.

Table 9-1. Number of bulblets formed after plant growth substance treatment versus water.

Cuttings

A cutting is a part of a leaf or stem that is cut from the "mother" or parent plant in order to make a new plant. A plant that has grown from a cutting is identical to the parent plant. See Table 9-2 for examples of plants successfully rooted from cuttings. Nursery plant growers usually use rooting compounds such as Rootone to help plants produce roots more quickly.

Activity: Propagating from Cuttings

Materials Needed

- Plastic garbage bag
- Scissors
- Sphagnum moss
- Ten plant cuttings (Chrysanthemums or similar)
- Rooting compound
- Tape
- Clear plastic bags

Procedure

1. Cut a plastic garbage bag into 8 × 12-inch strips. You will need two strips. Put a handful of moist sphagnum moss down on two 8 × 12-inch strips.
2. Get ten 2- to 4-inch-long cuttings of plants to be rooted. See Table 9-2. Chrysanthemums work well.
3. Take off any leaves on the bottom half of the stem. Dip the stems in rooting compound and tap off any excess powder.
4. Put the stems on top of the sphagnum moss so that the leafy ends extend over the edge of the bag 2 to 3 inches.
5. Fold the bag in half lengthwise, so that the ends to be rooted are covered with the bag. Roll the bag up and label it "treated".
6. Repeat the procedure for the other five stems, but do not dip them in the rooting hormone. Tape the bag with a piece of tape labeled "control".
7. Place each labeled roll in a clear plastic bag to keep the cuttings moist.
8. Check for signs of rooting after two weeks.
9. Transplant the cuttings when the roots are 2 inches long.
10. Keep a progress record in Table 9-3.

Plant Common Name	Botanical Name
Azalea	*Rhododendron spp.*
Boxwood	*Buxus sempervirens)*
Butterfly Bush	*Buddleia spp.*
Chrysanthemum	*Chrysanthemum morifolim*
Clematis	*Clematis spp.*
Cotoneaster	*Cotoneaster spp.*
Firethorn	*Pyracantha*
Forsythia	*Forsythia x. intermedia*
Fuchsia	*Fuchsia spp.*
Hydrangea	*Hydrangea quercifolia*
Jasmine	*Jasminum spp.*
Lavender	*Lavandula angustifolia*
Lilac	*Syringa spp.*
Rose	*Rosa spp.*
Spirea	*Spiraea spp.*
Verbena	*Verbena x. hybrida*
Viburnum	*Viburnum spp.*

Table 9-2. Plants successfully rooted from cuttings.

Stems	Week 1	Week 2	Week 3	Week 4	Week 5	Week 6
Control						
Treatment						

Table 9-3. Number and length of roots present weekly.

Layering

Layering is a very useful technique because it produces large plants for sale much more quickly than other methods such as cuttings. Rooting is stimulated when the flow of nutrients to the end of the stem is restricted by bending, cutting, or twisting copper wire around the stem. See Table 9-4 for successfully layered plants.

Activity: Layering Method

Materials Needed

- Three stems suitable for layering
- A sharp knife
- Wire
- Metal staples or large rocks to hold stems in place

Plant Common Name	Botanical Name
Apple	*Malus sp.*
Common Camellia	*Camellia japonica*
Dogwood	*Cornus sp.*
Dwarf Boxwood	*Buxus sempervirens*
Flowering Almond	*Prunus sp.*
Japanese Flowering Quince	*Chaenomeles japonica cvs.*
Lilac	*Syringa vulgaris cvs.*
Linden	*Tilia sp.*
Maple	*Acer sp.*
Purple Giant Filbert	*Corylus maxima 'Purpurea'*
Rhododendron	*Rhododendron sp.*
Saucer Magnolia	*Magnolia x soulangiana cvs.*
Viburnum	*Viburnum sp.*

Table 9-4. Plants successfully rooted by layering.

Procedure

1. Trim three suitable stems of leaves and side shoots starting four inches below the tip of the stem and continuing to 24 inches below the tip of the stem.

   CAUTION:
   Be very careful when using a sharp knife to avoid cutting yourself.

2. Beginning 9 inches from the tip of one stem, cut away a 1-inch strip of bark all the way around the stem that is closest to the ground when the stem is bent downwards. See Figure 9-3.
3. About 9 inches from the tip of another stem, twist some wire around the stem. It should be twisted so that it is finger-tight, but not digging into the bark.
4. Leave the third stem alone to serve as a control.
5. Dig a trench about 6 inches wide and 4 to 6 inches deep in the soil under each stem.
6. Bend each stem sharply 9 inches from the end. Lay the stem down in the trench. Keep the stem in place with metal staples or a rock so that it will remain in the soil when the trench is filled in. Fill in the trench with soil, making sure the soil is pressed well. Water the trench gently.
7. Allow rooting to take place for four to eight weeks. After four weeks, carefully dig up the stems to check for root formation.
8. Keep a progress record in Table 9-5.

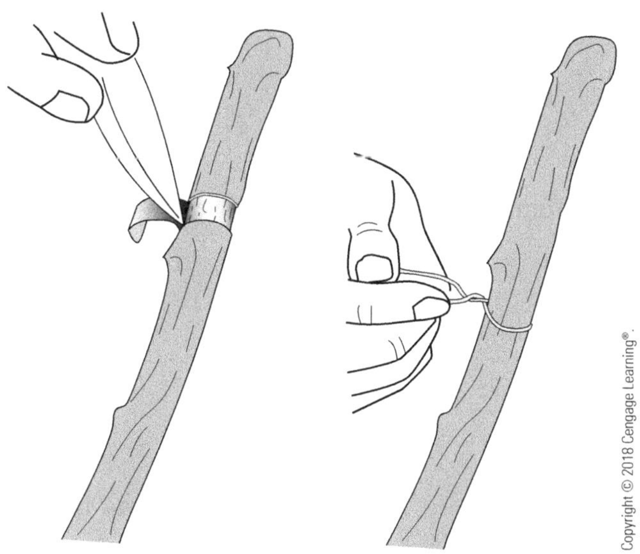

Figure 9-3. Girdling a stem.

Treatment	Week 2	Week 4	Week 8
Control			
Girdled			
Wired			

Table 9-5. Roots present on layered stems.

Questions for Thought

1. In which treatment of hyacinth bulbs did you see bulblets form the fastest?

2. Which cuttings rooted the fastest?

3. Which layered stem produced roots the fastest?

4. What advantages do plant nurseries gain by using asexual propagation to multiply landscape plants?

Glossary

asexual reproduction: producing more plants through some means other than seeds

bulblets: young bulbs

cambium: growing tissues that run parallel to the sides of roots and stems

grafting: process where two plants are joined together permanently by fusing the cambiums together

micropropagation: growing cells, tissues, organs, or parts of plants under sterile conditions for the purpose of asexually reproducing new plants

References

Browse, P.M. (1988). *Plant propagation.* New York: Simon & Schuster, 42–93.

Hill, L. (1985). *Secrets of plant propagation.* Pownal, VT: Garden Way Publishing, 57, 59.

MacDonald, B. (1986). *Practical woody plant propagation for nursery growers.* Portland, OR: Timber Press, 405–491.

Thompson, P. (1993). *The propagator's handbook.* North Pomfret, VT: Trafalgar Square Publishing, 55.

Wright, R.C., and Titchmarsh, A. (1987). *The complete book of plant propagation.* Great Britain: Ward Lock Limited, 73–118.

EXERCISE 10

Landscape Design

Student Objectives

After completing this lab activity, you should be able to:

- Understand the basic principles behind landscape design.
- Describe the processes of site analysis and maintenance.
- Construct a base map of a designated area and be able to form a master plan from the information provided in the base map.

Recommended Resources:

Read Chapters 6, 7, 8 and 18 in *Exploring Agriscience, 5th Edition.*

Introduction

Landscape design is defined as the arrangement of outside spaces for outdoor uses. The quiet gardens in town squares all across the nation are not there by chance. They are generally the results of a great deal of planning and a good basic design. **Landscaping** also has a functional side to it. Many garden plants can be grown beside show plants to create a beautiful and functional landscape design. It is necessary to do all the planning before you begin planting.

Analyzing the Site

A careful survey of the area should be made to determine whether the site can be used in the design or whether the design will have to be altered to fit the site. The character of the land, terrain, slope, and trees should determine the basic landscape pattern. The ground forms, existing vegetation, and the needs of the owner should determine how the land is developed. Every person has requirements and desires with regard to a landscape plan, so the area will be designed to satisfy the requirements of the people using the space.

Maintenance

Maintenance is an important consideration in landscape design. A low-maintenance plan is the goal of most homeowners. The area should be changed as little as possible. All attempts should be made to keep trees and other large natural features. The following are some helpful ideas for a low-maintenance design:

- Use ground covers or natural pine straw.
- Use fences or walls (see Figure 10-1) instead of clipped formal hedges for screening.
- Design raised flower beds for easy access and for easier weed control.
- Reduce size of **annual** flower beds. Use more **perennial** flowering trees and shrubs for color.
- Use native plant materials. (Find out which plants are native to your area and which of these grow best in the conditions that you require.)
- Keep the design simple.

Figure 10-1. Fences and walls can be used for screening eyesores and are lower maintenance than formal hedges.

Base Map

Preparing a base map is essential to developing a landscape plan (see Figure 10-2). From the base map, a plan for the entire lot is drawn. It is called the *master plan*. The master plan shows what the finished product will look like (see Figure 10-3). Mapping or plotting the area to scale from accurate measurements is an important step in the beginning of a landscape plan. Drawing the scale means letting 1 inch on paper represent a fixed number of feet on the ground. Use a scale where 1 inch on your paper equals 4 feet on the ground. To draw either plan, you will need an ordinary ruler and graph paper.

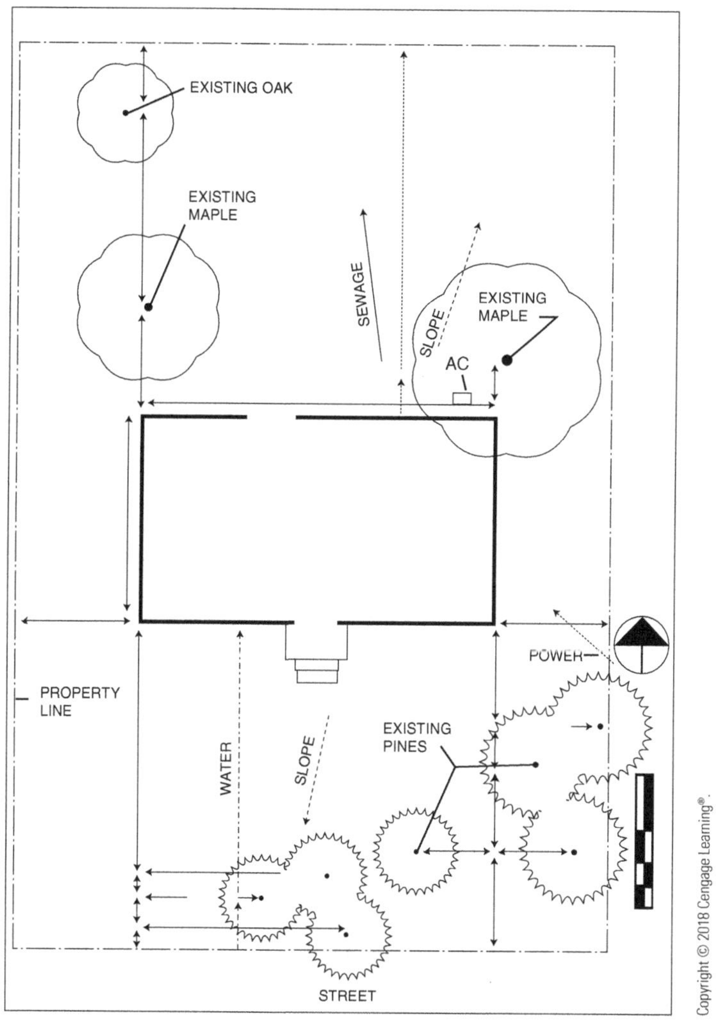

Figure 10-2. The base map shows the placement of the present greenery. Also, notice the scale and the arrow pointing north.

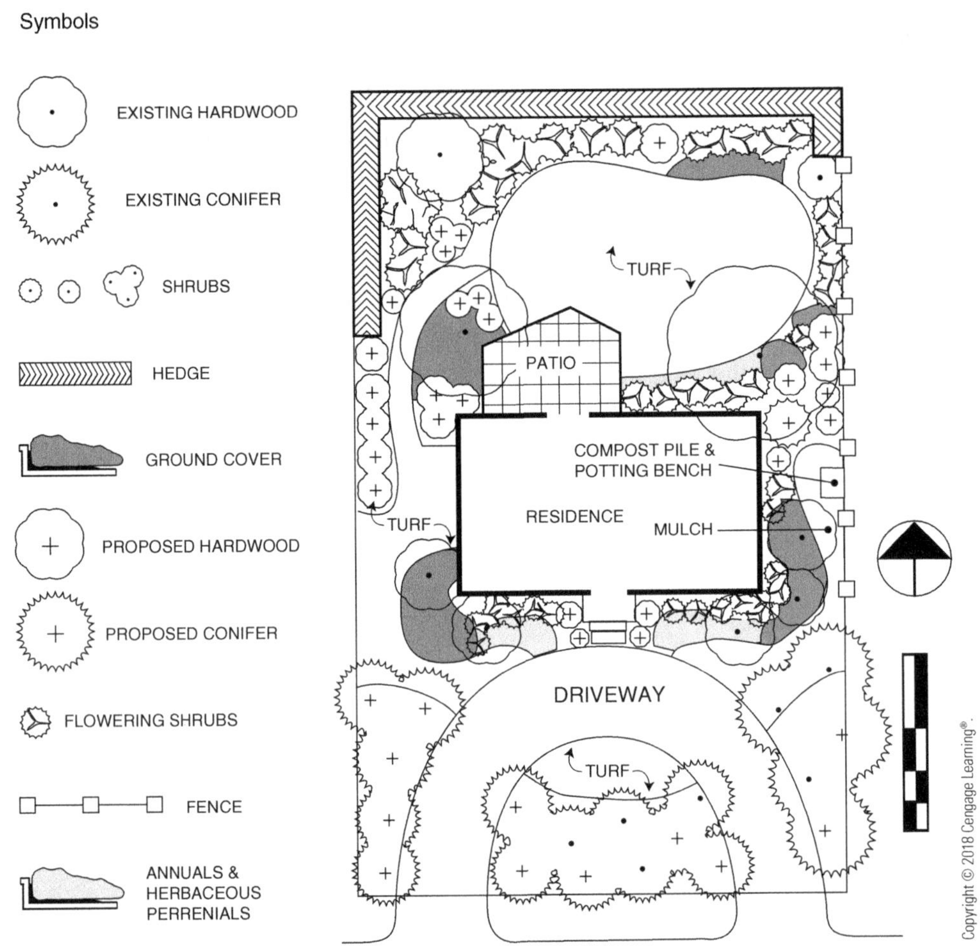

Figure 10-3. The Master Plan shows the finished product. Again, notice the scale and the arrow pointing north. The structures and areas are clearly labeled and there is a symbol list.

NOTE: The activities in this exercise suggest paper and pencil and other tangible tools to create these designs. However, if you have access to, and familiarity with, computer-based design software, that is also acceptable to use.

Activity: Drawing a Master Plan

Materials Needed

- Graph paper
- Plain paper for note taking
- A ruler
- A pencil

Procedure

Choose your home, the school building, or another familiar structure as your main building. Be sure to indicate your scale on your map.

1. Make a rough sketch resembling the general shape and proportions of the area.
2. Make a drawing of the main building correctly placed within the yard area. Be sure to show all doors and windows. Show the location of trees, shrubs, and flower beds.
3. Indicate location, length, and width of drive entrances, drives, walks, and other such features.
4. Above all, avoid cluttering the map!

Questions for Thought

1. What did you find the most challenging in drawing the master plan?

2. Is there anything that you would change in the master plan? Why/Why not?

Different Styles of Design

There are many styles of design. Figure 10-4 shows three examples of landscape design:

- The open lawn (10-4a), which is popular in the South.
- The naturalized setting (10-4b), where little, if any, of the natural site is disturbed.
- The private garden treatment (10-4c), frequently seen on urban lots. There is also a combination garden, which, as the name implies, is a combination of two or more of the other types.

10-4a. Open lawn

10-4b. Naturalized setting

10-4c. Private garden treatment

Figure 10-4. Three popular landscape designs. 10-4a is open lawn, 10-4b is naturalized setting, and 10-4c is the private garden treatment.

Activity: Preparing a Design

Materials Needed

- Graph paper
- Plain paper for note taking
- A ruler
- A pencil

Procedure

1. Look around your neighborhood at various current designs. Study them, then decide on one that you will create a design for. If possible, take a picture of the house.
2. Look at the picture and create a new landscape design for it. It will be necessary for you to make a drawing of the picture or to trace it if possible. Remember to follow maintenance and site criteria for the landscape design that you select.

Questions for Thought

1. Why did you choose the site that you did?

2. Describe the changes that you made in the landscape design and explain why you made those changes.

3. For all the sites, tell what style of design they have now; also tell what changes you might make to them to increase their beauty or functionality.

Activity: A Real Design

Materials Needed

- Graph paper
- Plain paper for note taking
- A ruler
- A pencil

Procedure

1. Construct a landscape design for the front of the school, the school sign, or another entrance.
2. Begin with a sketch of the structure. In this sketch include all the existing plants and structural pieces (sidewalks, fences, walls, and telephone poles).
3. Next, draw a base plan and formulate a master plan.
4. Ask for input from students, teachers, and school administrators.
5. After your master plan is complete, submit it to the school administration and see whether they will allow you to implement your design.
6. Remember to take into consideration all the important factors of landscape design.
7. It may be helpful to include a cost analysis in your proposal. The cost analysis is simply an item-by-item list of how much your materials will cost. You may be able to use clippings and grafts from local plants to lower the total cost.

Questions for Thought

1. List some of the things that you like about the present design.

2. Now, list some things that you feel need to be changed.

3. Sketch out the base plan and the master plan on graph paper.

4. Allow some of your classmates to review your design. What concerns did they have about your plan? Are these things that you are willing to change? Why or why not?

5. Why do you think you were asked to get some input from other students, teachers, and school administrators about what should be included in your design?

6. Look back at the base plan and the master plan that you made in the beginning of the lab. Are there things that you might change now that you are aware of the styles and principles of design?

Activity: Urban Garden Design

Urban gardens can be used to produce fruits and vegetables in a relatively small space. Raised beds are a popular choice for urban gardens because a high-quality soil can be used and harvesting is easier. For this activity, you will draw a design for an urban garden at your school using raised beds.

Materials Needed

- A measuring tape or measuring wheel
- Graph paper
- Plain paper for note taking
- A ruler
- A pencil

Procedure

1. Measure an area of the school that could be used for an urban garden. This could be a courtyard, a green space, or even a place that currently just has bushes or planting beds.
2. When choosing your site, take into consideration the amount of sunlight it gets. Notice whether the buildings block light for a significant portion of the day.
3. Label the paper with the scale you will be using.
4. Using your ruler and graph paper, outline the area that will contain the raised beds.
5. Next, draw the raised beds and label them with names of fruits, vegetables, and/or herbs that can be planted in each.
6. Create a price list for constructing and planting your design.

Questions for Thought

1. What are some benefits to using raised beds for planting fruits, vegetables, and herbs versus planting them in the ground?

2. Name at least three disadvantages to planting in raised beds.

3. Do you think the other students at your school would take more of an interest in where their food comes from and how it is produced if the raised beds were in an area more students would pass by?

Glossary

annuals: plants that bloom, mature their seeds, and die in one season

landscaping: an agricultural art used to make an area beautiful

landscape design: the arrangement of outside spaces for outdoor uses

maintenance: the act of keeping something clean and neat

perennials: plants that live longer than 2 years

EXERCISE

11

Insect Pollinators and Fruit Production

Student Objectives

After completing this lab activity, you should be able to:

- Describe flower parts and their interactions.
- List the steps in plant fertilization.
- Describe the role of pollination in fertilization.
- Give examples of pollinators.
- Describe the behavior of insect pollinators.
- Process fruit seeds according to accepted germination standards.
- Describe the process of grafting.
- Understand the effects of grafting on fruit production.

Recommended Resources:

Read Chapters 4 and 9 in *Exploring Agriscience, 5th Edition.*

Introduction

In flowering plants, the male and female reproductive cells develop within the flower (see Figure 11-1). Egg cells are produced in the **ovules**. One or more ovules are found inside the **ovaries**. **Sperm cells** are produced in the **anthers** and are carried by the **pollen grain**. **Pollination** must occur for a seed to develop. During pollination, pollen is transferred from the anther to the stigma of the flower.

After the pollen grain is deposited on the **stigma**, one of its cells produces a **pollen tube**. The pollen tube grows down the pistil and through an opening in an ovule until it finally reaches the **egg cell**. **Fertilization** occurs when a sperm cell travels through the pollen tube and joins with an egg cell. The fertilized egg cell develops into an embryo, which, if exposed to the proper conditions, will develop into a seed.

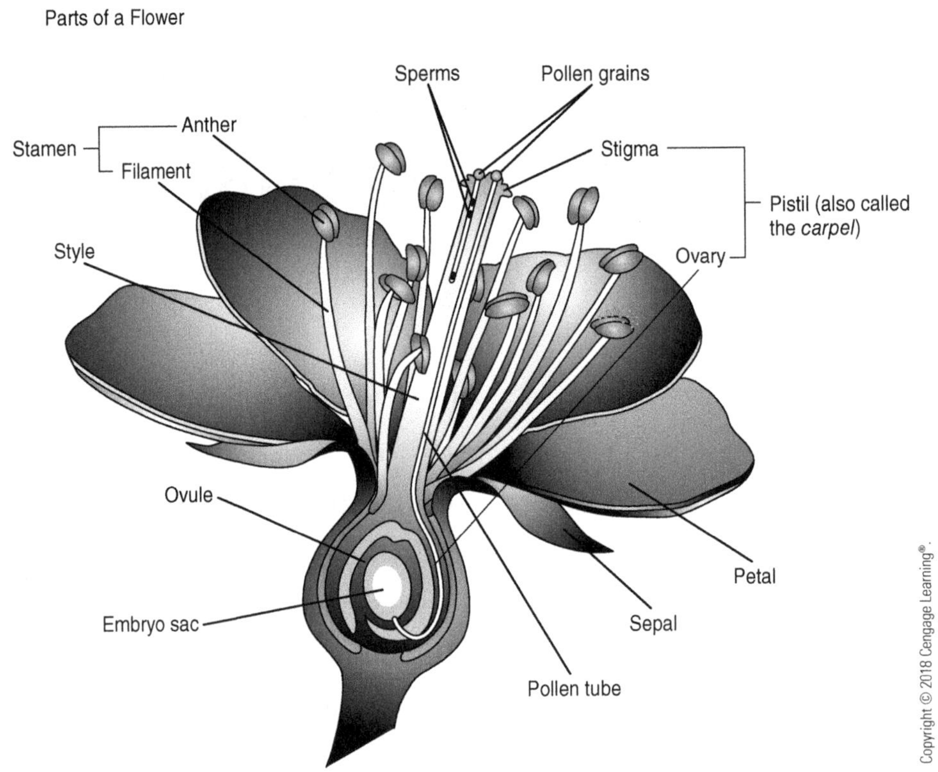

Figure 11-1. Illustrated flower parts.

Seed plants are divided into two groups, **gymnosperms** and **angiosperms**. Gymnosperms produce seeds that are uncovered. In the angiosperms, as the seeds begin to develop inside the ovary, the ovary matures into a fruit. The fruit helps protect and disperse the seeds. Seeds are protected from hungry insects, disease, and infection. **Seed dispersal** is the scattering or distribution of the seeds over a wide area. Dispersal can occur in several ways. Animals attracted to the fruit may carry it far away. Birds eat some seeds that pass through their digestive tracts and may be deposited far away. Wind and water can also carry fruits away from the parent plant.

The study of fruit and fruit production is called **pomology**. In order to produce fruits of good quality, it is necessary to understand how fruit plants grow and produce fruit. Fruit is important to our diet because it is low in fat and sodium and rich in vitamins A and C. It is also a good source of thiamin, niacin, iron, and calcium.

Pollination

Plants can be pollinated by the wind, insects, or animals. Flowering plants are usually pollinated by insects. Many grasses and trees are pollinated by the wind. Plants that are wind pollinated typically produce larger amounts of pollen to increase the likelihood of successful pollination. Flowers that are animal or insect pollinated usually have bright petals, nectar, and a sweet scent. As animals are attracted to the flowers, pollen grains stick to their legs and bodies. When the animal travels from flower to flower, some of the pollen grains are left behind. If the pollen is deposited in just the right place on the stigma, fertilization is possible.

Activity: Observing Insect Pollination

Materials Needed

- Access to a flower garden or flowing shrubs
- Colored pencils
- Paper

Procedure

1. When you arrive at the source of flowers, situate yourself approximately 6 feet away.
2. Sit quietly and observe the flower plot. Make a rough sketch of the plants and the area around them. Include as many individual flowers as possible in your drawing.
3. For the next five minutes, practice your observation skills and make notes of anything you see in the flower plot.
4. Observe the movement of an individual honeybee. Use a red colored pencil to trace the path the bee takes through the flower plot and the direction it takes when it leaves. Repeat this procedure with two more bees using orange and yellow pencils. Create a key at the bottom stating what the colors represent in your sketch.

   NOTE: Due to lack of availability, you may need to select insects other than those listed here.

5. Observe the movement of a single butterfly. Use a blue colored pencil to trace the path the butterfly takes through the flower plot and the direction it takes when it leaves. Repeat this procedure for two additional butterflies using green and purple colored pencils.
6. Observe the length of time each of the two insects remains on a single flower. You can do this by counting "one one thousand, two one thousand," and so on. Repeat the process three times for each of the insects. Record your observations in Table 11-1.
7. Observe the length of time each of the two insects spends between visits to flowers. You should start counting as soon as the insect takes flight and stop as soon as it lands on a different flower. Record your observations in Table 11-1.
8. Calculate the average number of seconds spent per flower and seconds between flowers for the different insects.

$$\frac{\text{Seconds/Flower Insect \#1} + \text{Seconds/Flower Insect \#2} + \text{Seconds/Flower Insect \#3}}{3} = \text{Average Seconds/Flower}$$

$$\frac{\text{Seconds between Flowers Insect \#1} + \text{Seconds between Flowers Insect \#2} + \text{Seconds between Flowers Insect \# 3}}{3} = \text{Average Seconds between Flowers}$$

Insect	Seconds Spent Per Flower	Seconds Between Flowers
Honeybee # 1		
Honeybee # 2		
Honeybee # 3		
Honeybee Average		
Butterfly # 1		
Butterfly # 2		
Butterfly # 3		
Butterfly Average		

Table 11-1. Timing the insects.

Questions for Thought

1. Write a paragraph about each of the insects you observed in the flower plot. You should include the average number of flowers each visited as well as time spent visiting flowers and between visits. In addition, you should include any other observations you made, such as the differences in the behavior of the insect species.

2. Do you think the behavior of the insects in the flower plot resulted in pollination? Honeybees are described as very efficient pollinators. Why do you think this is so?

3. Do you think plants that are insect pollinated or plants that are wind pollinated are more efficient at pollination? Why?

4. Are beneficial insects like the honeybee and the butterfly immune to the pesticides that are used to kill harmful insects? What do you think would happen if there were no butterflies or honeybees?

5. What do you think would happen if there were no pesticides? How might farmers find solutions to the problems in questions 4 and 5?

Germinating Fruit Seeds

Before fruit seed will germinate, a period of cool, moist storage at temperatures slightly above freezing is necessary. **Stratification** is one method used to condition the seed so it will germinate properly.

Activity: Fruit Seed Stratification

Materials Needed

- Fruit containing seeds
- Container of water for soaking seeds
- Peat moss
- Sand
- A plastic bag
- Access to cold storage
- Planting flat
- Potting soil

Procedure

1. Collect 25 fruit seeds.
2. Soak the seeds in water for 24 hours.
3. Prepare a stratification mix of 1/3 peat moss and 2/3 sand by volume.
4. Soak mix until it is thoroughly wet. Squeeze excess moisture out by hand.
5. Place the seeds into a bag and mix with the stratification mix.
6. Put bag in cold storage and hold there for the period indicated in Table 11-2 depending upon fruit seed used.

Fruit	Days
APPLE	070
PEAR	060
PEACH, NECTARINE	010
PERSIMMON	100

Table 11-2. Cold storage requirements.

7. Check the mix once a week to make sure that it does not dry out. Add water as necessary.
8. After the seeds have remained in the stratification mix for the required period, plant the seed in a flat of potting mix. Apple, pear, and persimmon seeds should be planted ¾ inch deep; peach and nectarine should be planted 1½ inches deep.
9. Water the flat and place in a warm area. Repeat watering as necessary.
10. Check the flats daily for seedling emergence and record your findings. Do this until it is evident that all the seeds have emerged that are going to.
11. Calculate the germination percentage for your fruit seeds.

$$\%\ \text{germination} = \frac{\#\ \text{of seeds emerged} \times 100}{\#\ \text{of seeds tested}}$$

Questions for Thought

1. A germination rate of 90 percent is desirable in fruit seeds. If your germination percentage was less, what are some possible reasons that could explain why the seeds did not germinate?

2. Why do you think the seeds need to be exposed to a moist, cold environment before they will germinate? (**HINT:** Think about fruit trees and their seeds as they evolved in the natural world.)

Producing a Fruit Tree

Unlike many other crops, commercial fruit plants are often not grown from seeds. Because seeds contain genetic information from two parent plants, there is no guarantee that the young plant will be exactly like either of the parents. A good analogy might be to families. A child's genes come exclusively from his or her mother and father, but this does not mean that the child will look or act *exactly* like either one of them. Likewise, a seed from two highly productive fruit plants might produce little fruit at all. The risk of having trees that produce little or low-quality fruit would be too great in planting an orchard or even restocking an existing one.

So where do fruit trees come from? They are produced through **asexual propagation**. In asexual propagation, new plants are produced through **cuttings**, **runners**, **layering**, **budding**, and **grafting**. Through all of these methods the new plants produced are *exactly* the same as the parent plant, and orchard owners can be assured that their new plants will produce quality fruit.

The peach growing industry makes use of grafting with the help of several different rootstocks. **Rootstocks** are the root system upon which named varieties of fruit have been grafted. For example, some apple varieties are grafted onto rootstocks that act to dwarf or control the size of the fruit. Lovell, Halford, and Nemaguard are the names of varieties of peach rootstocks. Seeds of the rootstocks are grown in a nursery for a full year. In about May, a bud of a desired variety is put on the seedling. After the bud starts growing, the seedling is cut off just above the bud. This allows the growth at the top of the peach tree to be from the bud, while the rootstock is from the seedling. In late fall or early winter, the grafted peach trees in the nursery are dug up and sold to peach growers to plant the following spring.

Activity: Grafting

Materials Needed

- Ten peach seeds
- Container of water for soaking seeds
- Peat moss
- Sand
- A plastic bag
- Access to cold storage
- Individual planting pots
- Complete fertilizer
- A sharp knife or budding knife
- Budding rubber or rubber bands

Procedure

1. Obtain about 10 seeds of Lovell, Halford, or Nemaguard.

   NOTE: If these varieties cannot be found, you can use any variety of peach seed. If peach seed is unavailable, other fruit seed can be substituted.

2. Stratify the seeds according to the steps outlined in "Fruit Seed Stratification" in this exercise.
3. Plant one seedling per pot using all the seedlings that germinated.
4. Wait until frost periods have passed and move the pots outside. If you have an early spring, you can wait to pot the seedlings at this time.
5. Keep the seedlings watered to encourage them to grow. It might be necessary to give them a little fertilizer (any complete fertilizer) every six weeks after potting. Fertilize by mixing 1 tablespoon of fertilizer per gallon of water and pour into the pots.
6. In late May or June, cut several new shoots 12 to 18 inches long from a quality peach tree. Keep the shoots in wet paper towels to keep them from drying out and try not to let them get too warm.

   CAUTION
   Be very careful when using a sharp knife to avoid cutting yourself.

7. With a sharp knife, preferably a budding knife, make a horizontal cut through the bark (but not into the wood) of one of the seedlings about 4 to 6 inches above the soil line. The horizontal cut should go about halfway around the seedling (see Figure 11-2).

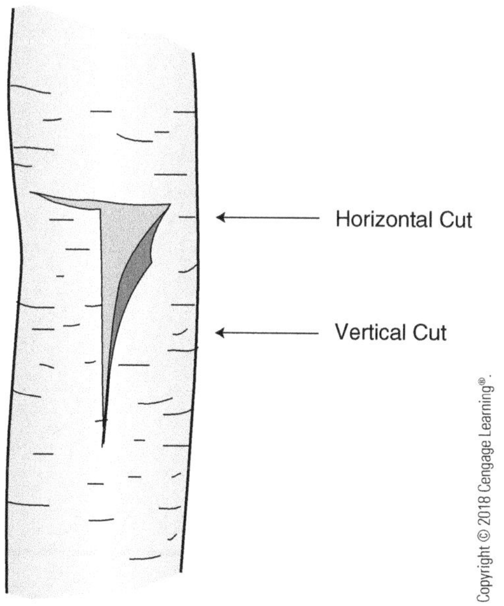

Figure 11-2. T-cut in seedling bark.

8. Next, make a cut straight down the seedling starting at the horizontal cut and continue for about 1 inch. Note that the bark can easily be peeled back from the seedling (as shown in Figure 11-2).
9. Cut a bud from the shoots you selected. Do not use the end or first couple of terminal buds. Cut off the leaf, leaving about 1/4 inch of leaf stem. A bud is at the axis of where the leaf joins the shoot. Cut just under the surface of the bud, starting about 1/2 inch above the bud and ending about 1/2 inch below the bud. It is not necessary to have the wood from the shoot, but it usually helps cut just into the wood to make sure you get the whole bud (see Figure 11-3).
10. Slide the freshly cut bud into the cut of the seedling. Be sure the bud is pointed upward. Start at the top of the T-cut and push the bud down into the leg of the T. The bark of the seedling should peel back easily (see Figure 11-4).
11. Tie the bud by wrapping it with a budding rubber or rubber band about 1/4 inch wide. Be sure to wrap above and below the bud and make it fairly tight. The last few wraps cross under the rubber to tie it off.
12. After about a week of growth, cut off the seedling just above the bud.
13. After the bud is growing well, rub off all leaves and shoots except those that are on the bud. Repeat as needed through growing season. About a month after the bud is growing strongly, cut the rubber band so it will not constrict growth. Now you have produced a grafted peach tree!

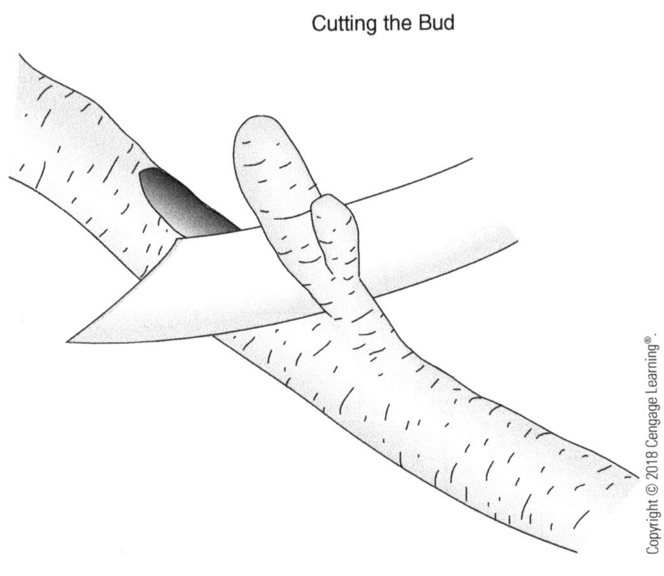

Figure 11-3. Excising the bud.

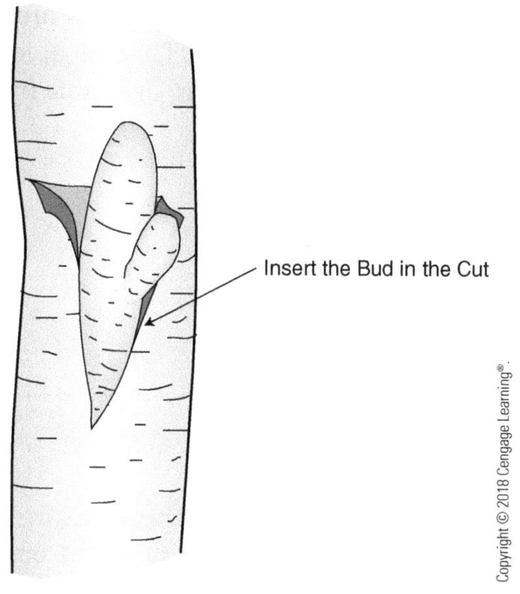

Figure 11-4. Inserting the seedling in the T-cut.

Questions for Thought

1. Why do you think the bud was able to grow in the cut of the bark? What procedures in the lab help the bud to begin growing on the seedling?

2. What are some possible reasons that a bud would not begin to grow on the seedling?

3. The example of using rootstocks in apple varieties mentioned above was directed towards producing smaller apples. Think of at least two different ways rootstocks might be used to produce more desirable fruits. Describe them.

Glossary

angiosperm: a type of seed plant in which the seeds are produced within the ovary that contain stored food for the growing plant embryo. Also called the flowering plants.

anther: found at the end of the stamen, produces pollen; male reproductive structure.

asexual propagation: producing new plants without fertilization

budding: a form of grafting where a bud from the parent plant is placed in a normal position on the cambium of the stock plant

cutting: any part of a plant that can be severed and is capable of regeneration

egg cell: the female reproductive cell, also called a germ cell

fertilization: the process in which the egg cell and the sperm cell unite to form an embryo

grafting: the inserting of a piece of one plant into another or the same plant with the intention that it will grow there

gymnosperms: a type of seed plant that produces naked or uncovered seeds

layering: the process of tying down a plant twig or shoot and partially covering it with soil so that it can take root without being severed from the parent plant

ovary: the part of the pistil that contains one or more ovules

ovule: tiny structures found inside the ovary of a flower. After fertilization, the ovule becomes the seed; female reproductive structure.

pollen grain: the male element of flowering plants that carries the sperm cell

pollen tube: grows from the pollen grain through the style and to the ovary. The sperm cell travels through the pollen tube to the egg cell inside the ovule.

pollination: the transfer of the pollen from the anther to the stigma, the first step in producing a fruit or a seed

pomology: the science of growing and handling fruits, especially fruit trees

rootstock: the root system upon which named varieties of fruit have been grafted

runner: an aboveground shoot that roots and forms young plants at some of the nodes

seed dispersal: scattering of the seeds away from the parent plant

sperm cell: the male reproductive cell, also called a germ cell

stigma: the receptive surface of the female organ that receives the pollen

stratification: the rest period that some seeds must have before they germinate. Generally, seeds must be exposed to chilling temperatures.

EXERCISE 12

Forest Plants and Land Use

Student Objectives

After completing this lab activity, you should be able to:

- Relate the science of ecology to forestry practices.
- Discuss the parts of a forest ecosystem in the United States including abiotic components and biotic components.
- Relate the variety of plant life in an area to forest management practices.
- Compute an index of similarity for two forest areas.

NOTE: This exercise discusses several concepts in ecology. You need to have a good working knowledge of math to complete this activity. A calculator is needed to finish this lab exercise.

CAUTION:
When working in forested areas, wear long pants and shoes that completely cover the feet (ankle boots are preferable in many areas). Always check for ticks when returning from any wooded area. If you use a chemical insect repellent, use the spray on your clothes, not your skin. You should change clothes as soon as you come back from the forest if you sprayed insect repellent on your clothing.

Recommended Resources:

Read Chapter 11 in *Exploring Agriscience, 5th Edition.*

Introduction

Ecology is the study of the interrelationships of organisms with one another and their environment. Ecology is a complicated subject that includes biology, chemistry, physics, geology, geography, and mathematics. Investigations by ecologists (scientists who study ecology) may concentrate on one organism, a group of organisms living in an area **community**, or on the organisms and their environment **ecosystem**.

Trees were removed from public and private forests in the United States between 1830 and 1891 at an alarming rate. Wood was needed to build railroads as settlers pushed westward, to rebuild the South after the American Civil War, and to build houses in the treeless prairies of the Great Plains. The land was also cleared of trees to produce crops for the growing population. The areas covered by forests in the eastern and northern United States decreased in size, and people began to realize that there might not be enough wood to meet demand.

In 1891, the U.S. Congress passed the Forest Reserve Act. Beginning with the Yellowstone Park Forest Reservation in 1891, the U.S. government became the caretaker of public forest lands in an effort to preserve and manage our forest resources.

Today there are increased demands on forested areas from industry. The demand for hardwood and paper products makes it necessary to manage forests wisely and use more parts of the tree. Books, newsprint, paper towels, toilet paper, disposable diapers, and paper bags are all made from forest products. Wood for construction and furniture manufacturing comes from forests. Many chemicals such as turpentine and resins are derived from commercially grown trees. Some trees provide food in addition to lumber. Conservationists are concerned about the possibility of large-scale tree cutting in old-growth forests, particularly in the northwestern United States.

Forest ecologists play an important role in the management of sensitive forest areas by providing information about how to keep a forest healthy. As we begin to move toward a time when forests will be sustained by good management practices, ecologists help the people making decisions about our forests to manage public and private forest lands wisely.

Forest Ecology

Forest ecology involves the study of the forest ecosystem. An ecosystem includes both the living communities of plants and animals and the physical environment in which they live. Ecosystems are made of two distinct components, the **abiotic** and **biotic** factors. Abiotic factors include all aspects of the nonliving environment such as rain and snowfall patterns, soil type, latitude, altitude, amount of sunlight, and temperature. Biotic factors are the living parts of the ecosystem: the plants, animals, protists, fungi, and bacteria found in the ecosystem. Forestry practices that affect the quality of the abiotic part of a forest ecosystem (such as fertilizers or the application of pesticides and herbicides) will also have an effect on the organisms in the forest. The removal of members of the forest community will also have an impact on the nonliving (abiotic) part of the forest ecosystem.

Forests are communities of organisms in which the dominant form of plant life is a tree. Forests require a great deal of water, and the types of trees in a forest are determined by the average air temperature. The animal communities in each forest are connected to the numbers and kinds of plants in the forest. As the number of different kinds of plants increases, the number of animal, fungal, and bacterial

species also increases. The tropical rain forest is the most complicated ecosystem on earth. The greatest **diversity** of living organisms is found in tropical rain forests.

Energy moves through the forest ecosystem, unlike nutrients, which cycle. The initial source of energy for the forest is sunlight. The forest plants, trees, shrubs, and herbs use sunlight energy to convert carbon dioxide (CO_2) and water to sugar molecules (glucose) through a process called photosynthesis. Photosynthetic plants and one-celled organisms are self-feeders; that is, they produce their own food. Self-feeding organisms are **autotrophs**. Autotrophic organisms are called **producers**. In the forest ecosystem, the producers are generally trees, shrubs, and plants growing on the forest floor.

Animals feeding on the plants use the plant sugars as an energy source. Organisms that eat living plant material are called **primary consumers**. **Consumer** organisms cannot make their own food; they must eat food produced by plant or plantlike one-celled organisms. Primary consumers are herbivores, animals that eat plants. Animals feeding on the animals that eat the plants indirectly use the plant sugars incorporated in the bodies of their animal prey for energy. Animals that eat primary consumers are called **secondary consumers** (see Figure 12-1). Animals that eat secondary consumers are **tertiary consumers**. Organisms that get food from outside their bodies (or cells) are called **heterotrophs**.

Questions for Thought

1. You survey two different forests—a natural forest that has oak, hickory, chestnut, beech, and elm trees and an area planted entirely with white pine trees. In which forest would you expect to find the greater animal diversity? Why?

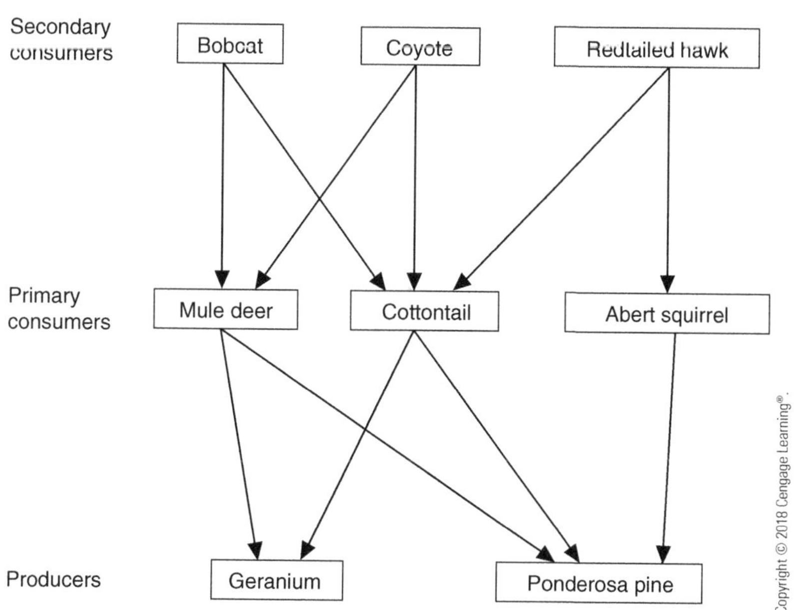

Figure 12-1 A simple food web with feeding levels.

2. There are three tree-harvesting methods: selective removal of individual trees, strip cutting (a row or several rows of trees are removed), and clear cutting (all the trees are removed). Which would be most economically efficient? Explain your answer.

3. Which method of harvesting would have the least impact on the forest ecosystem?

4. Your family owns 500 acres of land that is 80 percent forested. Timber was harvested from 150 acres 40 years ago. The area was replanted with three different kinds of trees. Fifty additional acres of timber were harvested over the past five years. Some parts of this area were replanted with pines. About 200 acres of the forest have not been cut in the last 200 years. You have been offered a large amount of money for the timber in the oldest hardwood stands. You wonder what will happen to the forest if the older trees are removed. Your family calls a forester and a conservationist to assist them in making a decision. What would you do and why?

Comparing Two Forest Areas: An Index of Similarity

The index or coefficient of similarity is an estimate of the extent to which two areas share the same species or individuals. There is more than one method for calculating similarity coefficients, but most foresters use the equation

$$Cs = 2jN/(aN + bN)$$

where a is the number of each kind of plant in Habitat a, b is the number of each kind of plant in Habitat b, and j is the number of plants shared by both habitats.

N is the number of plants counted. To see how this is calculated, let's look at a practical example:

Species	Maple	Red Oak	Shagbark Hickory	River Birch	Hemlock	
Forest 1	17*	42	28	4	0	91 = Na
Forest 2	22	19*	20*	0	15	76 = Nb

The lowest number for shared plant species is marked with an asterisk (*). Each forest area has at least that many individuals of that kind of plant. The marked values are added together to calculate *jN,* which equals (17 + 19 + 20) or 56. The coefficient of similarity would be (2 × 56)/(91 + 76) or 112/167, which is 0.6707. Forest habitats 1 and 2 have 67.07 percent of individuals and plant species in common. The closer the coefficient value is to one, the more alike the two forest habitats are. If the value is close to zero, the forest areas are very different. The index of similarity is used by foresters to identify forested areas that are very similar. Because the plant community determines the diversity of the animals in the forest, forested areas with high indexes of similarity would most likely have similar kinds of animals.

Activity: Forest Similarity

Materials Needed

- Clothing appropriate for going into the forest
- A field notebook
- Pruning shears
- Plastic bags for collecting specimens

Procedure

You and your class will go to two different forest sites selected by your teacher. Each group of students will walk a transect line marked by your teacher. You will need to wear appropriate clothing for this outdoor activity.

1. As you walk your transect, count the number of different kinds of trees.
2. You may want to remove a small twig with two or three leaves from each tree on your transect line (do not take large limbs). Put your twigs in a collection bag.
3. Record the number of each kind of tree on your transect line. If you are not certain whether two trees are the same, compare the leaf shape, arrangement, and bark pattern on twigs from each tree. You do not have to identify each species of tree. Simply record them as tree 1, tree 2, and so on.

4. Each group of students should walk a transect in each area (this can be done on two different days). If you collect twigs, look at them on the day they are collected or store them in the refrigerator or a cooler until you have time to study your samples.
5. Working with a group of three or four other students, analyze the twigs your group collected from both transect sites. Record the information in Table 12-1.
6. Calculate the coefficient (index) of similarity:

$$Cs = 2jN/(aN + bN)$$

where *a* is the number of each kind of tree in Forest Area 1, *b* is the number of each kind of tree in Forest Area 2, *j* is the number of species shared by both areas, and *N* is the number of trees counted.

TREE	Forest Area 1	Forest Area 2
Tree 1		
Tree 2		
Tree 3		
Tree 4		
Tree 5		
Tree 6		
Tree 7		
Tree 8		
Tree 9		
Tree 10		
Tree 11		
Tree 12		
ADD the columns	aN	bN
*Mark the lowest number that appears in a row. Add all of the marked (*) values to get jN = ________ Index of Similarity ________		

Table 12-1. Numbers of trees from transects through two different forest sites.

Questions for Thought

1. How similar were the two forested areas you studied?

2. If the two forested areas your class looked at are similar, are the similarities a result of the way that the land has been used? Explain your answer.

3. Could the similarities in the forests be a result of factors such as the age of the forest, the kind of soil, or similar patterns of temperature and rainfall? Explain your answer.

4. Thinking about the information you collected about these two forested areas, what is the connection between species diversity and forestry management practices?

Glossary

abiotic: the nonliving parts of an environment, such as the kind of soil and temperature and rainfall patterns

autotroph: an organism that uses chemical energy (such as sunlight) to make its own food, such as plants and algae

biotic: the living part of the environment; all of the organisms living in the ecosystem.

community: populations of different organisms living together in a common location

consumer: heterotrophs are also termed "consumers" because they consume energy-containing molecules that are either directly or indirectly made by the producers

diversity: a description of the relative number of different kinds of organisms living in a given area

ecology: the study of the interrelationships of organisms with one another and their environment

ecosystem: the level of ecological study or organization that includes all of the living organisms in an area and the physical environment in which they interact; the living community and the physical environment.

heterotroph: organisms that cannot make their own food and so must ingest energy-containing molecules produced by other organisms

primary consumer: an organism that eats producers such as plants or photosynthetic protists (also called a herbivore)

producers: autotrophs are also termed "producers" because they make food for themselves and other organisms

secondary consumer: an organism that eats primary consumers (also called a carnivore or meat eater)

species: a group of organisms living in the same location that can mate and produce fertile offspring

tertiary consumer: an organism that eats carnivores (secondary consumers)

EXERCISE

13

Organic Production of Vegetables

Student Objectives

After completing this lab activity, you should be able to:

- List the advantages and limitations of organic gardening.
- Describe insect control in organic gardening.
- Construct a small organic garden for growing vegetables.
- Determine the differences in growing responses of a variety of vegetables in different soils.

Recommended Resources:

Read Chapters 3, 5, and 13 in *Exploring Agriscience, 5th Edition.*

Introduction

Organic gardening uses natural minerals and organic fertilizers. Organic farmers grow their plants without the use of **synthetic pesticides** or **inorganic fertilizers**. The organic farmers keep the soil moist and fertilize with organic matter such as manure and compost. The soil is usually under cultivation at all times to prevent erosion. The plants are also rotated to reduce pest problems. One main advantage of organic gardening is the protection it provides to the environment. In addition, the organic farmer saves money by lowering the amount of water needed and reducing or eliminating the expense of synthetic pesticides or inorganic fertilizers. One of the main disadvantages is that an organic garden is time-consuming in cultivation and planning. There are some ways to lessen the chances of insects eating the plants. These include wrapping the plant with aluminum foil, putting mesh netting over plants such as cabbages, and planting certain plants such as herbs near the other plants to repel insects.

In this experiment, you create a small organic garden. You can either use the **compost** that you made in

Exercise 7 or you can make a supply of compost from the materials presented in this exercise.

Activity: Creating an Organic Garden

Materials Needed

- Compost from Exercise 7 or create a new batch
- Shredded plant materials (cut grass, dead leaves, etc.)
- Garden soil
- A pitchfork
- A cover
- A hotbed thermometer
- A shovel or spade
- A measuring tape
- String
- Plants that will produce quickly (carrots, leaf lettuce, radishes, spinach, peas, cabbage, cauliflower, and broccoli are suggestions)
- Fertilizer
- Graph paper

Procedure

1. List the ingredients that you included in your organic garden in the beginning of your journal.
2. Start the compost pile with a layer of brown material. Add some layers of green. Continue until you have three layers. If the greens are not fresh, add cottonseed meal to provide a nitrogen source.
3. Turn the pile every day to make sure that there is plenty of oxygen present. Take the temperature. If the pile does not heat up, add water or more nitrogen. When the material looks dark and crumbly, you have compost.
4. Keep a daily journal of your activities in making the pile. Describe your observations of how the compost pile is changing.
5. Dig out a 4' × 4' plot of earth to a depth of 12 inches with a spade. The class will be divided into two groups.
6. The first group will use the organic compost that was made to grow the plants. The rest of the class will use regular fertilizer and soil to grow their plants. Neither group should walk on the plot after getting it ready for planting.
7. Mark off 16 sq ft with string. The following are suggestions for planting for a garden that can be quickly produced: carrots, leaf lettuce, radishes, spinach, peas, cabbage, cauliflower, and broccoli. See Figure 13-1.
8. Sketch your organic garden on a separate sheet of paper.

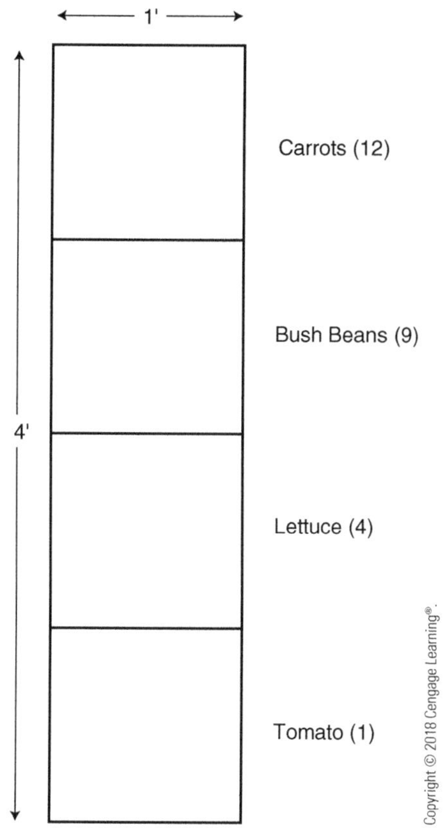

Figure 13-1. Sketch of a small organic garden. Listed are suggestions for planting as well as numbers of plants.

9. Record growth every other day and keep a journal of observations of your garden.
10. Every week, compare the growth rates of the two gardens.
11. Plot on graph paper the two gardens using time versus growth. Which grew faster?
12. Take a sample of the leaves from your garden. Draw them on a piece of graph paper. Did you see any loss due to insects?

Questions for Thought

1. What observations are made from the comparison?

2. What do you think is the reason for the differences that you observed?

3. Is there a difference between your group's foliage loss due to insects and the other group's loss of foliage? Why?

4. Are there any insects present? If so, are they damaging insects or not? How do you know this?

5. If there are damaging insects, what can be done to rid your garden of the pests?

Glossary

compost: a mixture of vegetable matter, used for fertilizing land.

inorganic: does not contain carbon and has never been living.

fertilizer: a preparation containing the elements needed for growth.

organic: contains carbon, has been living or is presently living.

pesticide: an insecticide used for destroying pests.

synthetic: artificial, made by humans.

EXERCISE

14

Animal Reproduction

Student Objectives

After completing this lab activity, you should be able to:

- Describe the process of gametogenesis including the similarities and differences between sperm formation and egg formation.
- Describe the processes of fertilization and cleavage.
- Name the three embryonic germ layers and describe the tissues and structures that develop from each.
- Distinguish between the four extraembryonic membranes in a vertebrate embryo and give the function of each in a bird egg.

Recommended Resources:

Read Chapter 14 in *Exploring Agriscience, 5th Edition.*

Introduction

Successful reproduction is essential to the maintenance of life on our planet. All organisms have some reproductive mechanism. One-celled organisms can reproduce by dividing to form two new cells. Reproduction by **mitotic** cell division produces two organisms that are genetically identical to one another and to the parent cell from which they are derived. This is an example of asexual reproduction. Asexual reproduction is common in plants, fungi, protists, and bacteria. Some simple animals such as sponges and jellyfish are also capable of asexual reproduction.

Higher animals such as vertebrates (animals that have a bony spine) cannot reproduce asexually. Sexual reproduction involves the union of two cells from different individuals resulting in offspring that possess unique genetic combinations of maternal and paternal characteristics. The sex cells that unite to form the new individual are produced by a special type of cell division, **meiosis.** Meiotic cell division results in four new **haploid** cells, each with half the number of chromosomes found in the parent cell. This reduction in chromosome number is necessary—the new offspring resulting from egg and sperm fusion should have the same total number of chromosomes as either parent. In this lab activity, you observe some of the events leading to the adult stage. You do not learn all the details, but you do learn about the principles of development.

Activity: The 2-day Embryo

Materials Needed

- Prepared slides of a 2-day-old chick embryo
- Dissecting microscope

Procedure

See Figure 14-1.

1. Observe how the embryo is twisted on its right side and its head bends like a question mark in reverse.

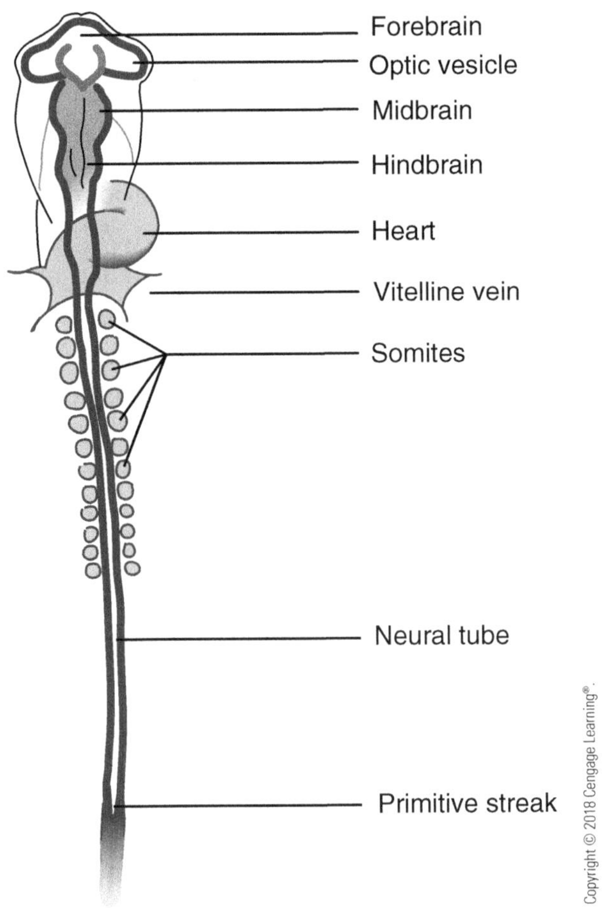

Figure 14-1. A dorsal view of a 33-hour chicken embryo.

2. Find the optic and otic vesicles, visible on either side of the brain. These will form the eyes and inner ear, respectively.
3. The heart sends blood forward through three pairs of aortic arches, and then posteriorly through the dorsal aorta toward the tail to the **yolk sac**. Many vessels branch over this sac, absorbing nutrients from it and acquiring oxygen before returning blood via the vitelline veins for another cycle through the heart and the embryo.
4. The chick embryo should contain pairs of **somites**. Can you count them on either side of the spinal cord? Begin at the hindbrain and count toward the tail. No limb buds are visible at this time.

Activity: The 4-day Embryo

Materials Needed

- A 4-day-old embryo still in the shell
- Dissecting microscope
- Small bowl (finger bowl)
- Scissors
- Forceps
- Spoon
- Watch glass
- Saline
- Paper towels

Procedure

Eggs at this stage contain so much yolk that it is difficult to view the embryo. Therefore, you will remove the embryo from the surrounding membranes to observe it. Obtain an egg and follow these instructions to open it (see Figure 14-2).

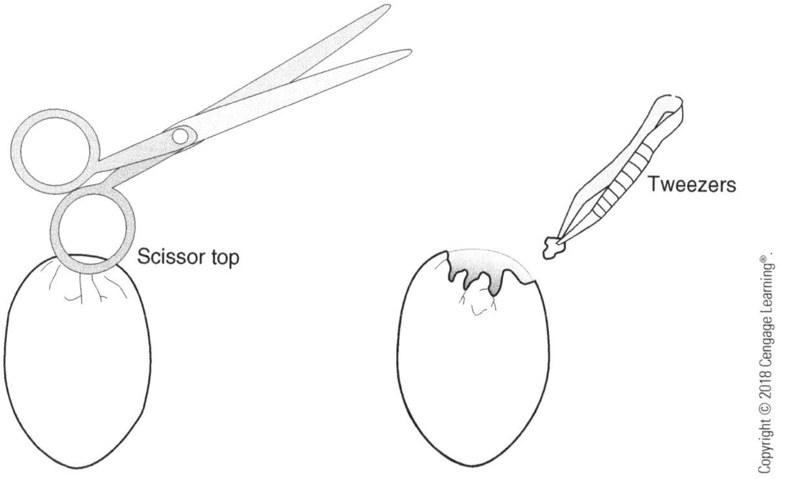

Figure 14-2. Procedure for opening a chicken egg.

1. Make a "nest" for the egg with a paper towel in a finger bowl so that the egg rests securely in the bowl. Place the egg in the orientation, as shown in Figure 14-2. Allow it to remain there at least two minutes before proceeding.
2. Crack the large end of the egg with the handle of the scissors.
3. Carefully and slowly clip the shell completely around the egg as indicated, and use the forceps to even the edges of the shell.
4. Remove the embryo from the yolk and place it in a dish of physiological saline for further examination. To do this, snip the area of the disk around the embryo and then move a spoon under it. Use scissors to separate the embryo from the albumen and the yolk. Then pour the embryo gently into the watch glass containing the saline solution.

Locate in a live 4-day embryo the structures, as labeled in Figure 14-3. Note that by the time the chick has undergone 96 hours of incubation, the following major events have occurred:

a. The embryo lies with its entire left side toward the yolk.
b. The wing and leg buds are now reasonably well-developed structures.
c. Between the hind limb buds, the **allantois** projects ventrally. It is one of the four extraembryonic membranes and functions in waste removal.

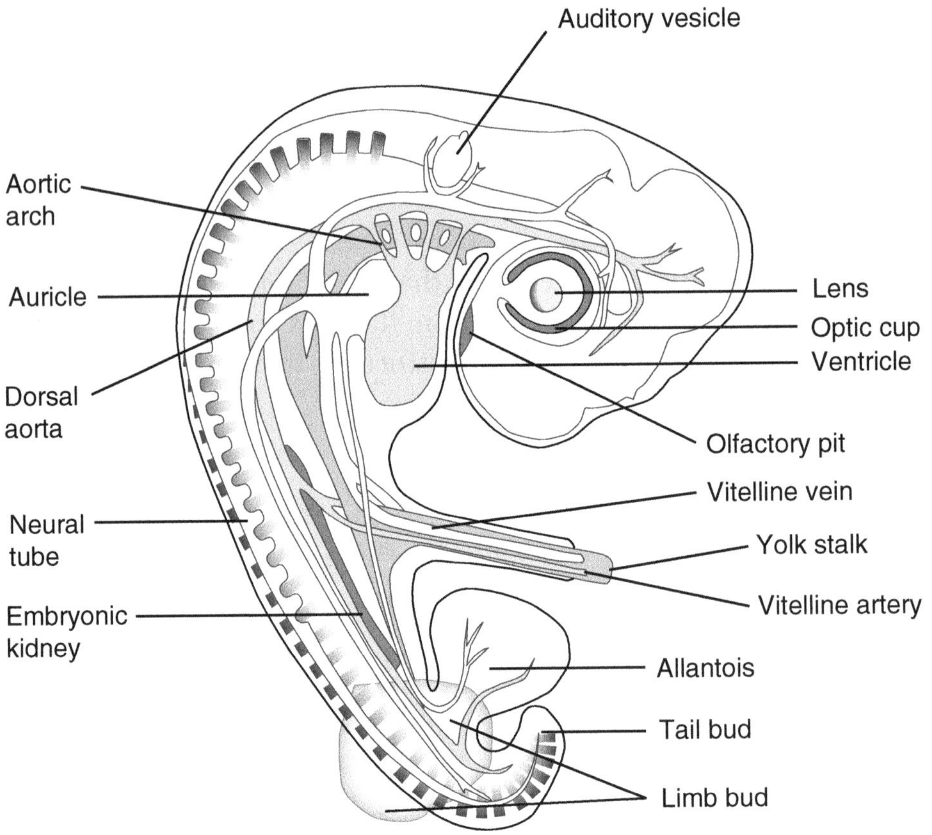

Figure 14-3. A side view of a 96-hour chicken embryo.

d. The **chorion**, formed at the same time as the **amnion**, serves with the allantois in the transport and exchange of respiratory gases. Note in Figure 14-3 that the allantois expands beneath the chorion. Differentiation continues until (and after) the egg hatches.

e. The **notochord** may be present along the dorsal surface and designates this as vertebrate. It provides support during skeletal development until it is replaced by the bony vertebral column.

Activity: The 8-day Embryo

Materials Needed

- An 8-day-old embryo still in the shell
- Dissecting microscope
- Small bowl (finger bowl)
- Scissors
- Forceps
- Spoon
- Watch glass
- Saline
- Paper towels

Procedure

1. Make a "nest" for the egg with paper towel in a finger bowl so that the egg rests securely in the bowl. Place the egg in the nest, in the orientation shown in Figure 14-2. Allow it to remain in that position for at least two minutes before proceeding.
2. Use a dissecting needle or probe to make a small hole in the shell or crack the large end of the egg with scissors or a scalpel handle.
3. Use forceps or fingers to pick away the shell. Initially avoid breaking the shell membrane and observe its position adjacent to the shell. Then carefully puncture the membrane.
4. Gently pour the entire embryo out into the finger bowl of warm physiological saline.
5. The chick can be seen within the chorio-allantois. Remove the membrane for further examination.

The following structures should be observed:

a. The eyes are large and conspicuous.

b. The beak and toes are distinct.

c. "Fingers" are appearing on the wing bud.

Questions for Thought

1. Look at a prepared slide of a 2-day chick embryo under a dissecting microscope. Draw and label the following structures: optic vesicle, otic vesicle, heart, neural tube, and somites.

2. Draw in detail the 4-day chick embryo. Label eye, brain, heart, and allantois.

3. Use the space provided to make comparisons of the 2-day, 4-day, and 8-day embryos.

   <u>2-day</u> <u>4-day</u> <u>8-day</u>

4. As eggs develop, the volume of yolk decreases relative to the volume of the embryo. Why do you think this is so?

5. What parts of bird eggs help them survive when laid in nests on the ground?

6. List the four extraembryonic membranes of a chicken egg and state the function of each.

Glossary

allantois: an embryonic membrane in the amniote egg that serves as a respiratory surface and as a waste storage organ; in placental mammals, the allantois forms most of the umbilical cord, playing an important role in placental development.

amnion: in amniote eggs and placental mammals, the amnion is the innermost extraembryonic membrane forming a protective fluid-filled sac surrounding the developing embryo.

chorion: the outermost extraembryonic membrane in amniote eggs and placental mammals; in the amniote egg, the chorion serves as a site for gas exchange and prevents the desiccation of the egg contents; in placental mammals, the chorion develops into the embryonic portion of the placenta.

haploid: possessing one set of chromosomes (N or 1N); formed following meiosis of diploid (2N) cells; in animals, sex cells are haploid; in plants, characteristic of the gametophyte stage, and some protists and fungi are also typically haploid.

meiosis: a type of cell division in which the chromosome number is reduced by one-half; one of each pair of homologous chromosomes passes to each daughter cell; usually occurs in gamete-producing cells.

mitosis: a type of cell division in which each daughter cell is genetically identical to the parent cell; the chromosome number remains constant in mitotic cell division.

notochord: in chordates (members of the phylum Chordata), the notochord is a dorsal flexible cartilaginous rod that extends the length of the body and provides support as a primitive skeleton; the notochord is present at some stage of development in all chordates; in vertebrates, the notochord is present only in the developing embryo and is replaced by a bony vertebral column.

somites: segments of mesodermal tissue that form during the differentiation of a vertebrate embryo.

yolk sac: an extraembryonic membrane in the amniote egg and placental mammals; in the amniote egg, the yolk sac surrounds the egg yolk; in placental mammals, the yolk sac is empty and forms a portion of the umbilical cord.

EXERCISE 15

Microbiology of Milk and Milk Products

Student Objectives

After completing this lab activity, you should be able to:

- Discuss the sources of undesirable microorganisms that contaminate milk products.
- Describe the effects of desirable microorganisms on milk products.
- Define pasteurization.
- Demonstrate how the process of pasteurization controls bacterial growth.
- Observe the preservative effects of culturing, salting, and refrigerating milk products.
- Describe the nature of bacterial growth in milk products.
- Define and give examples of cultured milk products.
- Describe flavor and aroma differences between cultured and noncultured milk products.
- Make a cultured dairy product.
- Define the major components of milk.
- Describe the process of homogenization.

Recommended Resources:

Read Chapter 15 in *Exploring Agriscience, 5th Edition.*

Introduction

M**ammals**, including humans, feed their offspring with milk produced in the mother's body. Most of our milk comes from cows, but we also use milk from horses, goats, buffalo, camels, reindeer, and yaks. The dairy industry calls fresh untreated milk from a cow **raw milk**. Before modern pasteurization techniques were developed, raw milk was a source of bacteria that caused infectious diseases. At the time milk is collected from mammals, it already contains bacteria. Milk can be contaminated by additional bacteria from a variety of sources, such as the hay and manure in dairy barns or improperly cleaned milk collection equipment.

The taste and smell of milk can be negatively affected when certain **microorganisms** increase in number to the point where their waste products are noticeable. The flavor and aroma sours and the milk **coagulates**, or lumps together, when large numbers of lactose-fermenting bacteria are present.

Public health agencies working in conjunction with the United States Department of Agriculture (USDA) have set sanitary standards for dairy products in the marketplace. Milk and other dairy products are subjected to pasteurization to control the number and growth of bacteria. During **pasteurization**, the raw milk is usually heated to 16°F for 15 seconds, destroying harmful microorganisms. This process ensures that the milk is safe to drink and also increases the amount of time that the milk will stay fresh.

Activity: Pasteurization of Pond Water

Although many of the microorganisms in pond water are much larger than the bacteria that contaminate milk, this activity simulates how the process of pasteurization sterilizes milk products.

Materials Needed

- Compound microscope
- Pond water
- Eyedropper
- Clean microscope slides and coverslips
- Hotplate
- 50 ml beaker
- Thermometer
- Tongs or heat-protective gloves

Procedure

1. Use an eyedropper to place a drop of pond water on a clean microscope slide. Place a coverslip on the drop of water.
2. Carefully examine the water under the microscope. Be sure to note the level of activity and movement of the microorganisms on the slide.
3. Turn on a hotplate and begin heating a small amount of pond water in a 50 ml beaker.

CAUTION:
When heating with the hotplate, be careful to keep clothing and books a safe distance away. Do not touch the hotplate with your hands.

4. Using a thermometer, note when the water reaches a temperature of 161°F. After the water has reached the desired temperature, make sure that it remains there for at least 15 seconds.
5. Using a pair of tongs or heat-protective gloves, remove the beaker from the hotplate and set it aside to cool. Turn the hotplate off.
6. When the beaker is cool to the touch, use an eyedropper to place a drop of the heat-treated water on a clean microscope slide. Place a coverslip on the drop of water.
7. Carefully examine the water under the microscope. Note the movement and activity level of the microorganisms.

Questions for Thought

1. Describe the differences you observed between the first sample of pond water and the heat-treated sample. What happened to the microorganisms?

2. People had the means to pasteurize milk and make it safe to drink long before they began to do it. Why do you think it took so long for people to realize the usefulness of pasteurization?

3. What are other ways foods are treated today to prevent the growth of bacteria and other organisms?

Activity: Controlling Bacterial Growth

Though pasteurizing milk products greatly reduces the number of bacteria found in them, the dairy industry employs several other methods of controlling bacteria. Cultured dairy products contain higher amounts of lactic acid, which lowers their pH and inhibits bacterial growth. These products tend to have a much longer shelf life than regular milk products. However, noncultured dairy products provide a fertile environment for the growth of bacteria. Refrigeration, canning, drying, and salting of cultured and noncultured milk products are all used to help prevent bacterial growth.

Materials Needed

- Whole milk
- Buttermilk
- Salt
- Scale
- Glass jars or beakers
- Spoon for stirring
- Plastic wrap
- Rubber bands
- Masking tape
- Marker
- Access to a refrigerator

Procedure

1. You will prepare four milk samples to assess the preservative effects of salt, refrigeration and culturing. Each sample should contain about the same amount of milk.
2. A salt level of about 17 percent has been shown to prevent the growth of most bacteria. Place the first glass on the scale. Record the mass of the glass. Add your sample of whole milk to the glass and record the total mass of the milk and the glass. Subtract the mass of the glass from the total mass to find the mass of the milk. Record your data in Table 15-1.

Total Mass (Glass + Milk)	Mass of Milk	Mass of Glass (Total Mass – Glass)

Table 15-1. Mass of milk.

3. Use the mass of milk calculated in #2 to estimate the amount of salt you will need to add. The salt should equal 17 percent of the total milk mass.

_____ grams of milk × 0.17 5 _____ grams of salt needed

4. Add the estimated amount of salt to the milk and stir well.

NOTE: The milk should be at room temperature before adding salt.

5. Seal the top of the glass with plastic wrap and a rubber band and set it aside. Label the glass "SALT" with masking tape and a marker.
6. Prepare two additional glasses of untreated whole milk and seal with plastic wrap. Place the first in the refrigerator and label it accordingly. Label the other "CONTROL" and place it with the "SALT" glass. These samples should remain at room temperature.
7. Finally, prepare a sample of buttermilk and seal as described in step 5. Label the glass "BUTTERMILK" and place it with the "SALT" and "CONTROL" glasses at room temperature.
8. Observe the milk samples at four, six, and eight days. Record your observations in Tables 15-2 through 15-5. Indicate relative amounts of growth and odor with + signs (+ for little growth; +++ for moderate growth; and +++++ for extensive growth).

CAUTION:
Smell the milk by waving your hand over the glass toward your face. Do not smell the milk directly.

Time Elapsed	Odor	Growth
4 Days		
6 Days		
8 Days		

Table 15-2. Salt sample.

Time Elapsed	Odor	Growth
4 Days		
6 Days		
8 Days		

Table 15-3. Control sample.

Time Elapsed	Odor	Growth
4 Days		
6 Days		
8 Days		

Table 15-4. Refrigerator sample.

Time Elapsed	Odor	Growth
4 Days		
6 Days		
8 Days		

Table 15-5. Buttermilk sample.

Questions for Thought

1. How did the growth of bacteria compare among the four samples? Which do you think is the most effective method for controlling bacterial growth?

2. What do your results suggest about the nature of bacterial growth?

3. Make a hypothesis of how a sample of raw milk (unpasteurized) would have compared to the other samples after 8 days at room temperature. Explain your answer.

4. Before refrigeration, how do you think people dealt with the spoilage problems associated with dairy products?

Activity: Growing Bacteria

Controlling the growth of bacteria through pasteurization is extremely important in dairy production. Bacteria can multiply so rapidly that even if a few dangerous bacteria remain alive in dairy products, their numbers will grow very quickly. However, the rapid growth rate of bacteria can help us observe them in laboratory settings. If we provide a bacterium with good growing conditions, it can produce a group of new bacterial cells that is large enough for us to see in a relatively short time.

Materials Needed

- Raw milk
- Aluminum foil
- Wax paper
- Paper cups
- Potato (sliced)
- Dissecting probe
- Tape
- Marker

Procedure

Part A

1. Prepare two bacterial "growth chambers." For each chamber, cut out a circle of aluminum foil 12 centimeters in diameter and a circle of wax paper 8 centimeters in diameter. Cut off the top 3 centimeters of a paper cup.
2. Mold the foil around the top of the paper cup to form a small dish.
3. Cut a slice of potato and put it in your foil dish.
4. Cover the foil dish with the circle of wax paper, crimping the edges of the aluminum foil and wax paper together. Do not close the dish completely; leave one side open. Close the opening with tape and print your name on the tape.

NOTE: The growth chambers should be sterilized. Your teacher can sterilize these for you in an autoclave or pressure cooker for 10 minutes at 67.5 newtons (15 pounds per square inch).

Part B

5. Set one of the growth chambers aside. This chamber is the control. Label it "C," for control.
6. Open the taped end of your second growth chamber. Using the sterilized end of a dissecting probe, dip the tool into raw milk.

   Rub the probe on the potato, spreading a very small amount of raw milk. You should make two rubs on the potato slice, each about the size of a dime.

   CAUTION:
   If flaming is used to sterilize the probes, you should take extreme care not to burn yourself or your clothes. Your teacher may sterilize the tools for you.

7. Reseal the second chamber with tape after rubbing the potato with raw milk. Set this chamber aside with the first growth chamber.

Part C

8. Open both growth chambers after two days. Examine the potatoes for bacterial growth. Small white, cream, or yellow dots are colonies of bacteria resulting from one bacterium that has grown and multiplied very rapidly to form millions of bacterial cells.

   CAUTION:
   Do not touch the potato surface with your hands because it will result in the bacteria transferring to you. This could potentially cause you to be ill if accidentally inhaled or ingested.

9. Count the number of bacterial colonies on the surface of the treated potato. Compare this number with the number of colonies on the untreated (control) potato. If separate colonies cannot be counted, compare the surface of the untreated potato with the surface of the treated potato.

   CAUTION:
   Be sure to wash your hands thoroughly after handling the growth chambers. Dispose of the growth chambers by sealing them in a plastic bag and placing in the garbage container.

Questions for Thought

1. Describe the difference you found between the treated and untreated potatoes. What did you expect to find?

2. What are some possible explanations for your results? If you had unexpected results, what might have happened to contaminate your samples of potato? (**HINT:** Is the raw milk the only source of bacteria in the experiment?)

3. Bacteria need both food and water to survive and reproduce. What provided these to the bacteria in your experiment? Do you think dairy products might provide a good source of food and water for bacteria to grow?

4. If bacterial colonies grow so rapidly, why don't we see them everywhere? Why haven't they crowded out other forms of life?

5. Assume that under ideal conditions bacteria divide every 20 minutes, doubling their numbers. Starting with a single bacterium under these circumstances, how many would there be after 2 hours? Four hours? Eight hours?

Activity: Cultured Milk Products

Although there are many harmful types of bacteria that are not desirable in dairy products, there are also beneficial bacteria that are actually added to dairy products. In cultured dairy products, various bacteria are added to milk or milk products. Generally, cultured dairy foods have a characteristic tart flavor due to the conversion of some of the milk sugar (**lactose**) to lactic acid by the cultures. Buttermilk, sour cream, and yogurt are all examples of **cultured dairy products**.

CAUTION:

If you are allergic to dairy products, you should not participate in the tasting part of the lab. Work with a partner who can make the flavor observations for you. You can still describe the smell in your notes.

Materials Needed

- Samples of six different dairy products
- Small paper cups

Procedure

1. Your teacher will prepare six samples of dairy products in small paper cups for you and your partner to taste, three of which are cultured dairy products. The samples will be labeled only with the numbers 1 to 6.
2. Using a clean spoon for each sample, taste the six dairy products and record your description of their flavor and smell. Based on your observations, hypothesize whether the dairy product is cultured or not. Record your results in Table 15-6.
3. When everyone has finished tasting the samples and recording their observations, the teacher will share with you which of the samples are cultured products.

Sample Number	Description of Flavor and Smell	Cultured Products? Yes or No
1		
2		
3		
4		
5		
6		

Table 15-6. Growth and odor of sample.

Questions for Thought

1. Did you correctly predict which of the dairy samples were cultured products?

2. Describe how the cultured dairy products taste compared to the regular dairy products. Why do the cultured products taste this way?

3. Which of the two groups of dairy products do you prefer? Which did your classmates prefer? Do you think there is a larger market for regular or cultured products?

Activity: Making Sour Cream Butter

CAUTION:
If you are allergic to dairy products, you should not participate in the tasting part of the lab.

Materials Needed

- One quart of extra heavy whipping cream
- Two tablespoons of active culture buttermilk
- Clean plastic bottle or butter churn
- Cold (refrigerated) water
- Salt
- Plastic wrap
- A butter knife and sliced bread for spreading finished butter on

Procedure

1. To one quart of extra heavy whipping cream, add two tablespoons of active culture buttermilk and mix well.
2. Allow the mixture to sit overnight at room temperature.

   CAUTION:
   This can be done safely only if active culture buttermilk is added. The buttermilk produces acid, which will prevent spoilage from occurring.

3. Churn the mixture until you first see small pellets of butter—about 20 minutes. They should be about the size of the end of your little finger.

   NOTE: If you do not have a butter churn, you can place the liquid in a clean soda bottle and shake it vigorously.

4. Drain off the buttermilk.
5. Add refrigerated water to the butter pellets. Shake or stir the water to rinse the pellets. Drain the water off.
6. Dissolve a small amount of salt into one tablespoon of cold water. Add to the butter pellets.
7. Begin working the mixture with your hands until it forms into a single large glob. At this point, you can shape your butter any way you like, wrap it in plastic, and set it in the refrigerator to harden.

   NOTE: Working the butter will warm it up. If it gets too warm, it will begin melting and make a big mess. To prevent melting, put the mixture in the refrigerator periodically.

8. You can now enjoy your homemade sour cream butter on bread or crackers!

Questions for Thought

1. The sour cream butter you made is a cultured dairy product. Does it taste different than regular butter? How so? Why does it taste different?

   CAUTION:
   If you are allergic to dairy products, you should not participate in the tasting part of the lab. Work with a partner who can make the flavor observations for you.

2. Cultured dairy products require the action of certain types of bacteria. What ingredient in the butter recipe provided the bacteria? (**HINT:** You might read the labels of the ingredients.)

3. How did your butter spread on a slice of bread? Why might this be a problem in marketing butter?

Activity: Milk Separation

Milk is made up of a watery portion called **whey** and a fatty portion that contains **curd** and **cream**. The whey contains protein, sugar, and other soluble materials, and the curds and cream contain fat and minerals. Cheeses, which are typically high in fat, can be made from the curds. The presence of these fatty and watery portions of milk can cause problems with milk products. The fat and water will eventually separate from one another. Because the fat is less dense, it rises to the top of the milk in the form of cream, leaving the whey and curd on the bottom. However, the process of **homogenization** helps break up the globules of fat to a size small enough that they will remain suspended and not rise to the top after standing for a period of time.

Materials Needed

- A wide-mouthed jar
- One cup vinegar
- One-third cup oil

Procedure

1. In a wide-mouthed jar, combine 1 cup of vinegar and one-third cup of oil.
2. Mix vigorously for three minutes.
3. Let the contents settle and observe the results.

Questions for Thought

1. What do you think the two layers in the jar represent?

2. What do you think made the two layers separate? How do you think this happened?

3. Have you ever seen this phenomenon in the milk you have at home? Under what circumstances did it occur? What are some other characteristics of milk that has undergone this separation of fat and water?

Glossary

coagulate: transforming a liquid into a soft, semisolid, or solid mass

cream: the sweet fatty liquid that is separated from cow's milk; it contains more than 18 percent milk fat.

cultured dairy product: dairy products to which lactose-fermenting bacterial cultures are added to achieve desirable effects

curd: the coagulated part of milk that results when milk is clotted by rennet, by natural souring, or by the addition of a starter

homogenization: breaking up and dispersing the fat particles in milk to a size small enough that they will not rise as cream

lactose: milk sugar

mammal: the animal group in which the female produces milk and nurses the young

microorganism: an organism that is so small it cannot be seen clearly without the use of a microscope

pasteurization: any food-preserving process using heat that does not exceed 212°F

raw milk: fresh, untreated milk as it comes from the cow

whey: the watery portion of milk that remains after the curd and cream have been removed

EXERCISE 16

Chick Development and Production

Student Objectives

After completing this lab activity, you should be able to:

- Identify the main parts of the bird egg.
- Follow the stages of chick development from before egg-laying to hatching.
- Explain the development of the chick embryo through the 21st day.
- Care for and hatch fertilized chicken eggs.

NOTE: This lab contains some exercises where you will have to observe a fertilized chicken egg. If you are uncomfortable with this, ask your teacher if you are allowed to complete an alternative assignment. You should, however, be aware that the study of embryonic structures has greatly enhanced the progression of science in the past. Alternative assignments could include one of the following:

- Write a paper on the egg production and chicken industry in the United States.
- Construct clay models of the embryo at different developmental stages.
- Produce a video of the lab that could be used to supplement future labs or replace them altogether.

CAUTION:
The egg whites and the yolks can be contaminated with Salmonella *bacteria.* Salmonella *can cause diarrhea and severe cramping. Be sure to wash your hands thoroughly after handling chicken eggs and chicks.*

Recommended Resources:

Read Chapter 16 in *Exploring Agriscience, 5th Edition.*

Introduction to the Avian Egg

The avian (bird) egg is a wonder of nature. Eggs (chicken eggs, for example) are complex structures that begin as an **ovum** that is fertilized, forming a **zygote**, which grows and is eventually laid as an egg. Eggs need only a warm, humid environment while the embryo is maturing.

What is an egg?

The outside of the egg is called the shell. The shell is **porous**, and the pores at the large end are larger and more numerous than those at the small end. Under the shell are the outer and inner shell membranes, which protect the contents of the egg. The shell membranes surround and contain the white or **albumen** of the egg. The albumen is the main water source for the developing embryo (see Figure 16-1).

At the 20th day of **incubation** (development of the chick inside the egg) the chick pokes its beak into the air cell. By this time, the cell has become larger, and the chick draws its first breaths of air from this space. In a fresh egg you can see two white cords, the **chalazae**, attached to the yolk and to the inner shell membrane at the ends of the egg. The chalazae hold the **yolk** in the center of the egg. The yolk contains a large amount of fat as well as vitamins and minerals needed for normal growth. The fat in the yolk combines with oxygen that is taken in through the pores of the shell, and together they provide large amounts of energy. The **blastoderm** is really the "true egg." From the blastoderm the embryo is formed. The remaining parts of the egg are to feed, care for, and protect the developing chick embryo.

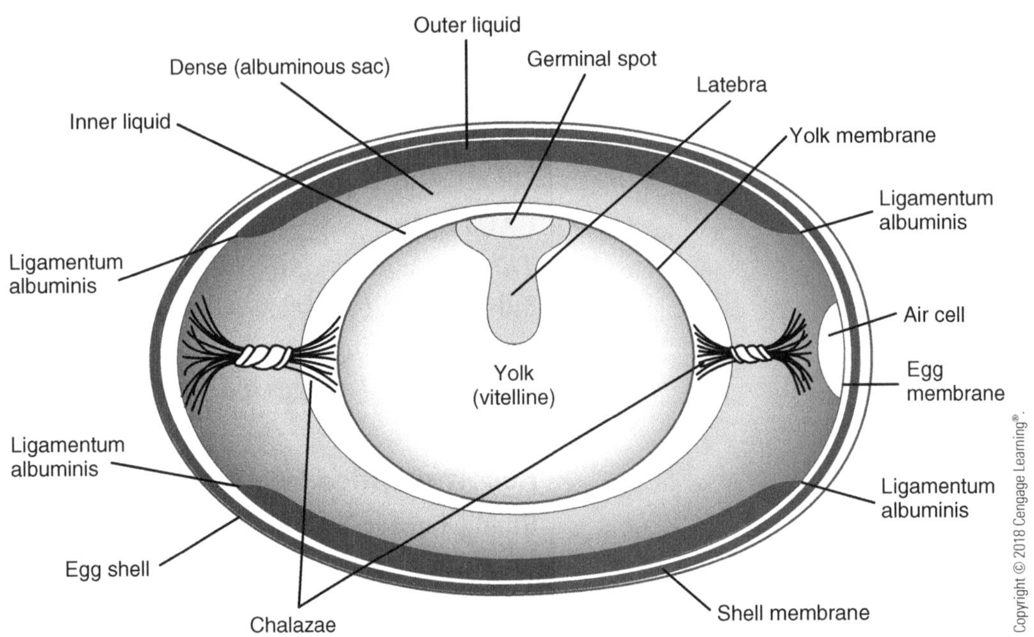

Figure 16-1. The internal anatomy of the avian (bird) egg.

Activity: Parts of an Avian Egg

Materials Needed

- Five raw (unfertilized) eggs
- Three small dishes
- A zip lock bag
- A very bright light

Procedure

1. Break an egg into a dish. Referring to the drawing of the egg (Figure 16-1), identify the following parts: inner and outer shell membrane (found inside the shell), chalazae chords, albumen or white of the egg, yolk, germinal disc (blastoderm).
2. Take an egg, put it into a zip lock bag, and squeeze it as hard as you can, applying equal pressure on all sides. Do this at arm's length over a bowl. If you apply equal pressure and the egg is not cracked, you can feel the egg flex in your hand, but it will not break.
3. With the help of your teacher, examine an egg that is three weeks old, one that is 10 days old, and one that is as fresh as possible. Hold them point down and shine a bright light through them. Notice the difference in the size of the air cells at the round end of the eggs. A fresh egg has an air cell slightly larger than a dime. As the liquid in the egg evaporates, the air cell becomes larger. Break the eggs into three separate dishes and observe the differences in them. As the egg ages, the egg white spreads out and the egg yolk becomes flatter (see Figure 16-2). Draw what you see in the space provided.

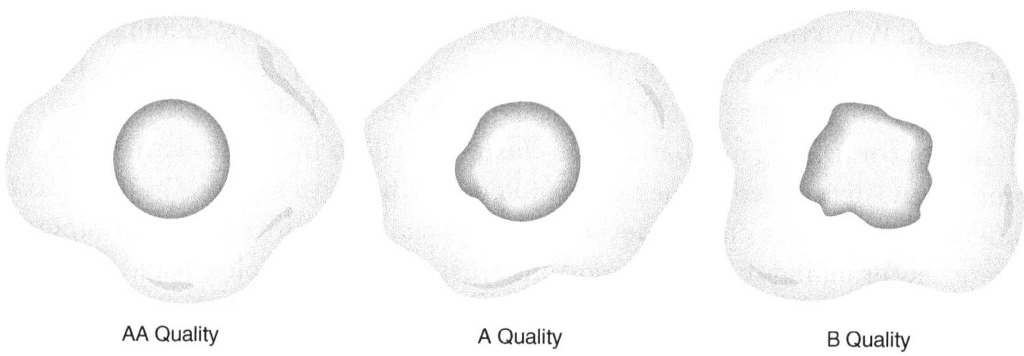

Figure 16-2. Grades of eggs. Notice that the fresher the egg, the more round the yolk.

Questions for Thought

1. What grade is your egg? (**HINT**: See Figure 16-2!)

2. See whether you can identify other animals that come from eggs. How are seeds and eggs similar?

Activity: From the Egg to the Chick

One of the greatest miracles of nature is the development of the egg into the chick. After 21 days of incubation, a fully developed chick comes from an egg. The development of the chick begins with a zygote. This cell is formed in the union of the male and female sex cells in the process known as **fertilization**. In birds, the fertilization occurs about 24 hours before the egg is laid.

Development Prior to Egg-Laying

The newly formed zygote begins to divide, first in 2, then in 4, 8, 16, 32, 64 cells, and so on. When the egg is laid and cooled, development stops. Cooling the egg to room temperature does not always result in the death of the embryo. It may start its development again after several days of rest if it is heated by the hen or an **incubator**. During incubation, the cluster of cells in the blastoderm begins to multiply. As division continues, some differences begin to appear. Gradually, the different parts of the cluster develop into special structure and cell groupings. These new groupings become important organs, and the whole cluster, made up of many millions of cells, becomes a new chicken (see Table 16-1).

Physiological Processes Within the Egg

Special temporary organs or membranes are formed within the egg to provide for its feeding (**yolk sac**), breathing (chorion), waste removal (**allantois**), and protection (amnion) (see Figure 16-3). These organs function within the egg only until the time of hatching and form no part of the fully developed chick.

Before egg-laying:	Fertilization Division and growth of living cells Separation of cells into groups of special function
Between laying and incubation:	No growth; stage of inactive embryonic life
During incubation:	
1st day:	
16 hours	First sign of likeness to a chick embryo
18 hours	Appearance of alimentary tract
20 hours	Appearance of vertebral column
21 hours	Beginning of formation of nervous system
22 hours	Beginning of formation of head
23 hours	Appearance of blood islands—vitelline circulation
24 hours	Beginning of formation of eye
2nd day:	
25 hours	Beginning of formation of heart
35 hours	Beginning of formation of ear
42 hours	Heart begins to beat
3rd day:	
50 hours	Beginning of formation of amnion
60 hours	Beginning of formation of nose
62 hours	Beginning of formation of legs
64 hours	Beginning of formation of wings
70 hours	Beginning of formation of allantois
4th day:	Beginning of formation of tongue
5th day:	Formation of reproductive organs and sex determination
6th day:	Beginning of formation of beak and egg tooth
8th day:	Beginning of formation of feathers
10th day:	Beginning of hardening of beak
13th day:	Appearance of scales and claws
14th day:	Embryo turns its head toward the blunt end of egg
16th day:	Scales, claws, and beak becoming firm and horny
17th day:	Beak turns toward air cell
19th day:	Yolk sac begins to enter body cavity
20th day:	Yolk sac completely drawn into body cavity. Embryo occupies practically all the space within the egg except air cell.
21st day:	Hatching of chick

Table 16-1. Important events in development of the chick embryo.

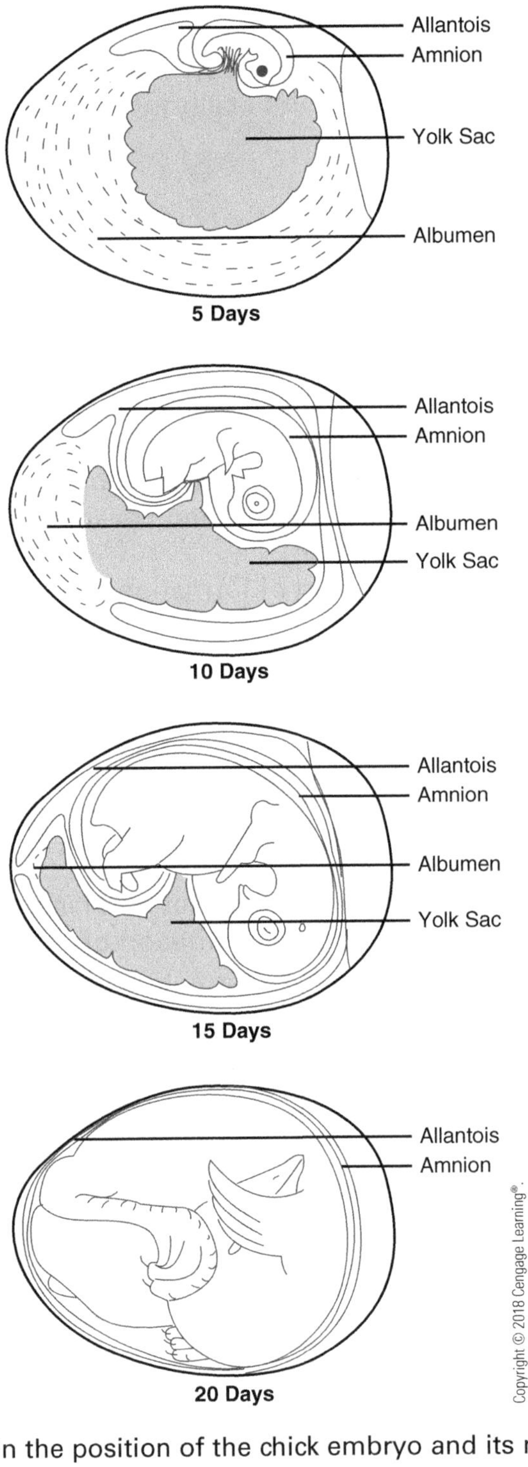

Figure 16-3. Changes in the position of the chick embryo and its membranes.

Hatching

The time of hatching is important in the life of a chick. On the 21st day the chick begins its escape from the egg. The first break in the shell is made by a sharp horny structure, the egg tooth, near the tip of the beak. The chick begins to breathe normally. Hatching is completed by the chick's slowly turning in the shell and chipping it in a circular path. Then, with a strong twist of its neck, the chick causes the walls of the shell to begin to shake and fall apart; freedom is gained. The chick is wet, but a few hours later the chick, now dry and fluffy, takes full advantage of its freedom and walks about its new and mysterious world. The usefulness of the egg tooth is over, and it will be lost in a few days.

Activity: Hatching Chicks

Materials Needed

- still-air incubator (obtain or built)
- pan with water
- fertile eggs
- thermometer
- marking pen
- separate cover for hatched chicks
- Water-soaked sponge (if necessary)
- Wet-bulb thermometer (if available)
- egg cartons (optional)

Procedure

1. Refer to Table 16-2 for information on incubating eggs.

   The still-air incubator should be operated with water in the pan for several hours before fertile eggs are set. During the warmup period, the temperature should be approximately 100°F (37.7°C). The temperature should be kept between 97°F and 103°F. If it stays at either 97°F or 103°F for several days, the hatch may not be as good as expected. The ideal temperature is 100°F.

   Fertile eggs can take a lot of abuse because they are so well insulated and protected, but they are very sensitive to heat. Running the incubator at 105°F for 30 minutes will greatly damage the embryo, but running it at 90°F for three or four hours will merely slow the growth rate.

Temp. °F	Wet-Bulb Reading in Still-Air Incubators					
100	81.3	83.3	85.3	87.3	89.0	90.7
101	82.2	84.2	86.2	88.2	90.0	91.7
102	83.0	85.0	87.0	89.0	91.0	92.7
Percent relative humidity	**45**	**50**	**55**	**60**	**65**	**70**

Table 16-2. Wet-bulb readings and their relationship to relative humidity.

2. Just before setting the eggs, mark the date on one side of each. This will serve as a record of the date on which the egg is set and also will show that the egg has been turned at turning time. Some people set eggs in egg cartons for easy turning, but for the best results lay the eggs directly on the screen in the incubator.
3. Turn the eggs at least three times each day; however, the more often the eggs are turned the better they will develop. Good results are obtained when eggs are turned first thing in the morning, again at noon or after school, and the last thing at night.

   Turn the eggs an odd number of times each day—that is, three, five, or seven times. If there is a long period of time between two of the daily turnings, the side that is up the longest will be staggered from day to day. The eggs do not need to be turned during the last three days of incubation.
4. Check the humidity periodically. Humidity in the incubator should be 50 percent the first 18 days and 65 percent the last three days. If you use a wet-bulb thermometer, you can determine the relative humidity from Table 16-3.

   An easy way to determine the amount of moisture in the incubator is to candle the eggs. As incubation progresses, the air cell of the egg becomes larger because the egg loses moisture. The normal size of the air cell at 7, 14, and 20 days is shown in Figure 16-4.

   Too much drying because of low humidity (large air cell) causes the chick to stick to the shell; not enough drying because of high humidity (small air cell) may drown the chick in the shell.

Requirements	Chicken/Bantam
Incubation Period (days)	21
Forced-Air Operating Temperature (°F, dry-bulb)	99.75
Humidity (°F, wet-bulb)	85–87
Do Not Turn Eggs After	19th day
Operating Temp. Last 3 Days of Incubation (°F, dry-bulb)	99
Humidity, Last 3 days of Incubation (°F, wet-bulb)	90–94
Open Ventilation Holes One-Fourth	10th day
Open Ventilation Holes Further if Needed to Control Temp.	18th day

Table 16-3. Incubator period and incubator operation for chicken and bantam eggs.

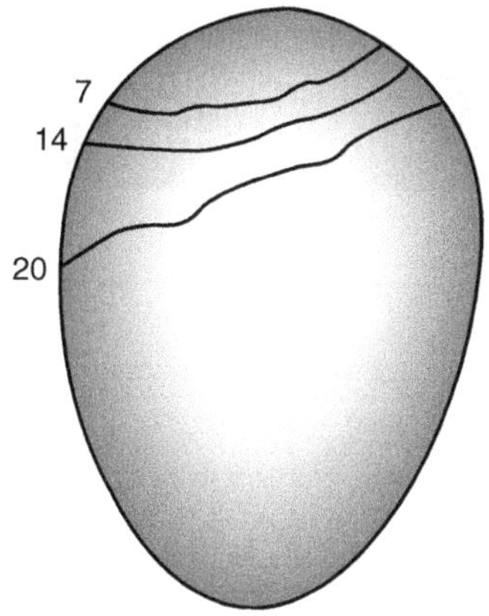

Figure 16-4. Showing the air cell at 7, 14, and 20 days of incubation.

5. If necessary, relative humidity can be raised by adding a water-soaked sponge to the incubator or by using a larger pan (more surface area). It can be lowered by decreasing the surface area of the water container or by opening the incubator.

6. Record the date, number of times the eggs were turned, the temperature of the room, and the temperature and relative humidity of the incubator in Table 16-4 each time the eggs are turned.
7. When the eggs hatch, remove the chicks as soon as they fluff up. Otherwise the incubator will become quite dirty.
8. Place the chicks under a cover with the temperature adjusted to 95°F and give them feed and water.

Questions for Thought

1. What is the function of the yolk sac? The allantois?

2. What structure does the amnion protect?

Date	Time Eggs Turned	Room Temp.	Incubator Temp.	Incubator Humidity	Remarks (Who turned eggs)

Table 16-4. Incubation Data Chart.

3. Label the parts in Figure 16-5 as best you can.

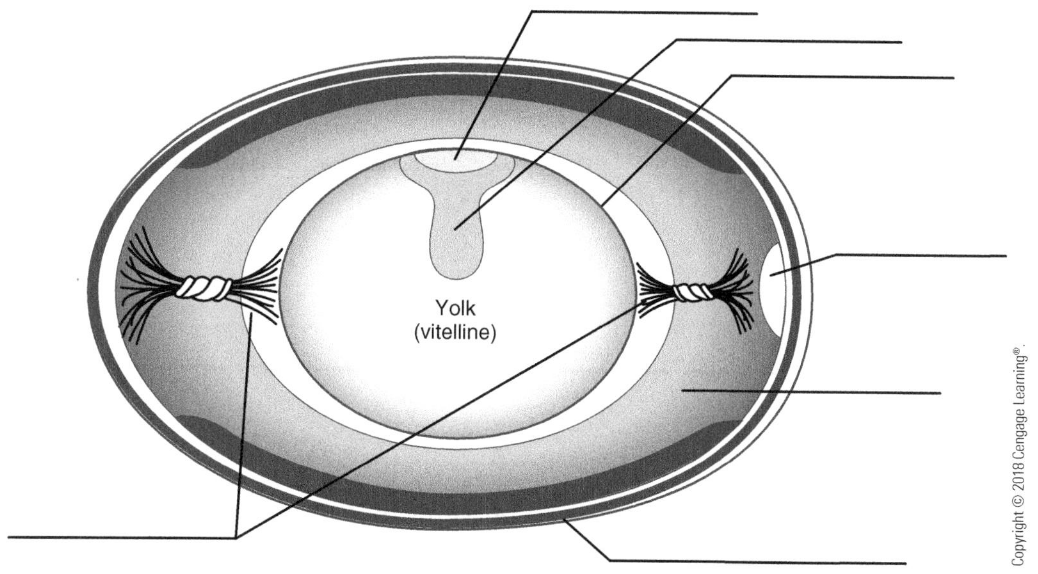

Figure 16-5. Label this diagram with the following terms: germinal spot, latebra, yolk membrane, chalazae, egg membrane, shell membrane, air cell.

Advanced Activity: The Incubator

NOTE: The following activity provides instructions on how to build your own incubator. If time allows, your teacher may choose to assign this as a class project. Due to the complexity of this activity, and the inherent hazards, the incubator should only be built under the close supervision of a teacher.

A simple, inexpensive still-air incubator can be constructed from a Styrofoam ice chest. The exact size of the ice chest to be used will depend on the number of eggs to be incubated at one time and the size of the chest available.

Heat

Heat is provided by two 40-watt lightbulbs mounted in porcelain sockets. The amount of heat is controlled by a wafer-type thermostat with a snap-action switch. This thermostat can be purchased, or one can be taken from an old electric brooder. One porcelain socket is mounted inside each end of the ice chest about 3 inches below the top edge. The thermostat is mounted on one side of the chest about 2 inches below the top edge with the water on the inside and the adjuster on the outside. Using large-diameter washers on the screws or bolts for mounting the porcelain sockets will help prevent breaking the Styrofoam.

The porcelain sockets are wired in parallel, not series, so that each light will burn independently of the other (see wiring sketch, Figure 16-6). All wire contacts should be covered for safety.

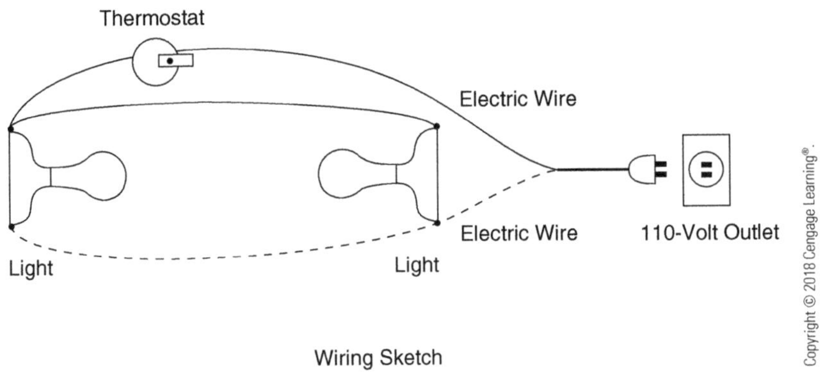

Figure 16-6. Wiring sketch.

Humidity

Humidity can be supplied from a pan of water placed in the bottom of the ice chest. A cake pan approximately 1½ inches deep is sufficient. The dimensions of the cake pan should be somewhat smaller than those of the ice chest so that it may be easily removed.

Egg Tray

A simple egg tray or platform can be made from 1/4-inch hardware cloth or welded wire. Cut a piece of hardware cloth so that its dimensions are 6 inches longer and 6 inches wider than the inside diameter of the bottom of the ice chest. Cut a 3-inch square out of each corner of the hardware cloth and bend the projecting pieces so that they form legs (see Figure 16-7). Trim the rough edges and cover them with tape so they will not puncture the Styrofoam. The platform should fit loosely over and around the water pan so that it may be easily removed. Do not allow room for chicks to get out of the tray into the water pan, as they will drown.

Ventilation

Ventilation can be provided through small holes in the sides of the ice chest. A sharp round instrument, such as a pencil, will be satisfactory for making the holes (see Figure 16-8). Twist the instrument and push gently through the Styrofoam, being careful not to break the sides of the chest. To prevent the Styrofoam flaking and filling the holes, heat a large nail and sear the surface. Make a total of 16 holes approximately 1/4 inch in diameter. On each side of the chest make four holes approximately 2 inches from the top and four holes approximately 3 inches from the bottom. Space the holes approximately 4 inches apart.

Window

A window is not necessary, but it will allow you to make observations without removing the top and causing a change in temperature and humidity. Place a

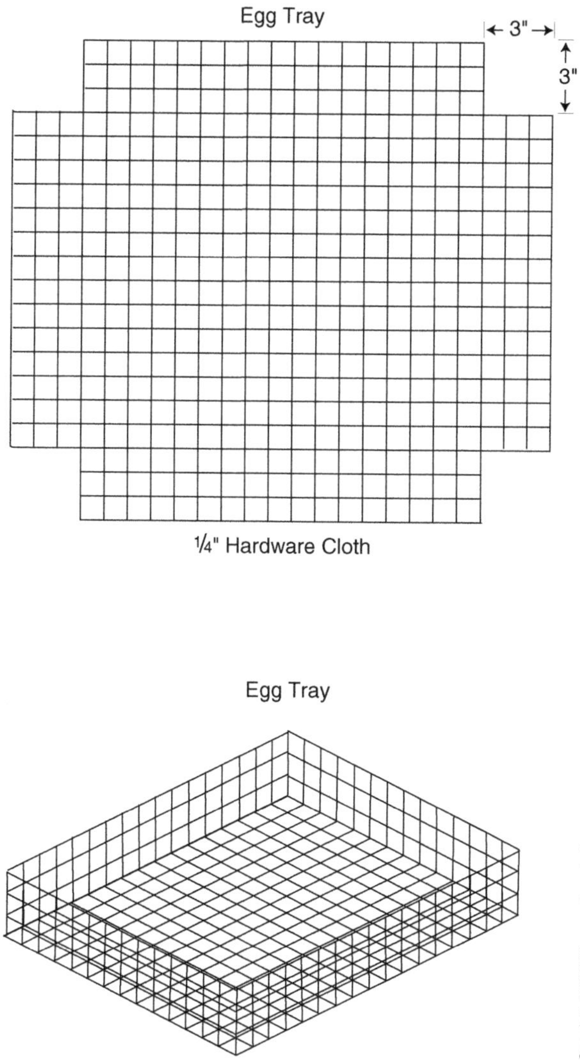

Figure 16-7. Cutting and folding the egg tray.

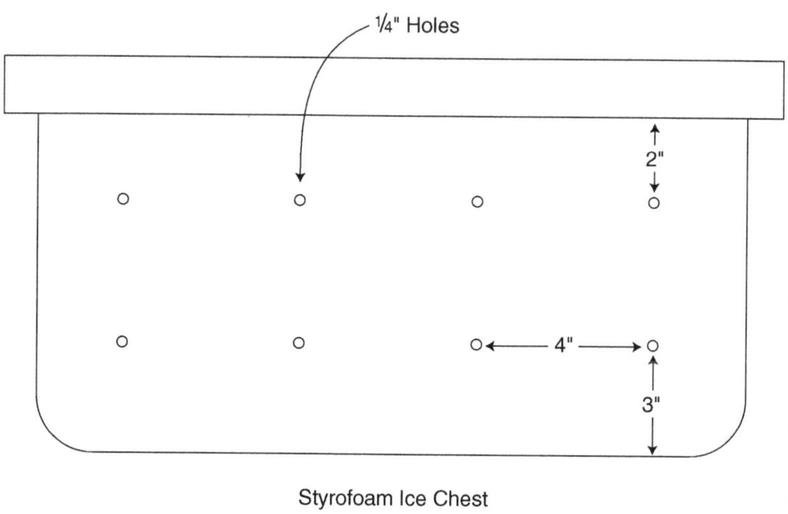

Figure 16-8. Cutting ventilation holes in the ice chest incubator.

piece of glass on the top of the ice chest. Make a cut ¼-inch deep around the outside edges of the glass. Do not cut through the top of the chest (see Figure 16-9). Remove the glass and draw a line 3/4 inch inside of the cuts. Following the line, cut all the way through the top of the chest and discard the piece that is cut out. Around the inside edges of the opening, draw a line 1/4 inch from the top of the chest. Cut along this line to a depth (about 3/4 inch) that meets the original cut and a strip can be lifted out. Place the glass in the recessed area and secure it with strips of tape around the edges of the glass.

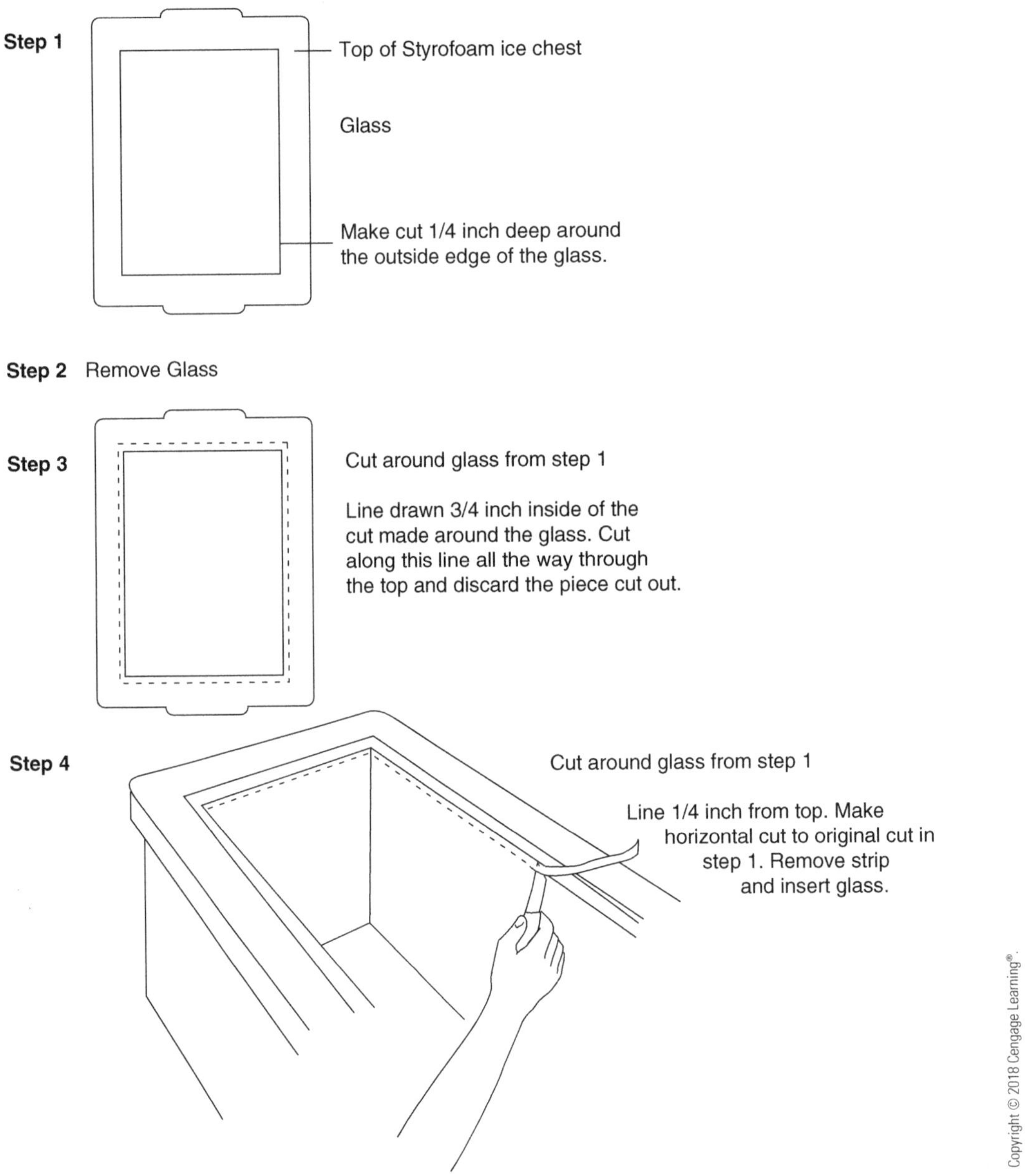

Figure 16-9. Putting in the glass cover.

Test the Incubator

The ice chest is now an incubator. Fill the water pan approximately half full with warm water and place it in the bottom of the incubator. Place an egg tray over the water pan. An incubator thermometer can be placed on the egg tray or attached to the inside of the incubator so that the bulb is about 1 inch above the egg tray. Place the top on the incubator and plug the lead cord into a 110-volt outlet. Adjust the temperature control bolt until the lights go on. Continue to adjust the temperature control bolt until the thermometer indicates the desired temperature.

Glossary

albumen: the protein portion of the egg, the egg whites; also the source of liquid for the embryo.

allantois: a sac that contains many blood vessels; it is the structure that allows for gas exchange between the embryo and the outside of the egg.

blastoderm: the "true egg." This structure, sometimes called the germinal disc, is the part of the inside of the egg that grows into the chick.

chalazae: twisted protein fibers that hold the yolk in place in the middle of the egg.

fertilization: the union of the sperm and the egg.

incubation: the process of warming eggs before they are hatched.

incubator: a structure that allows you to incubate eggs.

ovum: the female gamete or egg.

porous: having many holes.

yolk: the source of food for the growing embryo.

yolk sac: the structure that holds the yolk and supplies the embryo with food.

zygote: the cell that results from the joining of the egg and the sperm.

EXERCISE 17

Water Chemistry and Fish Production

Student Objectives

After completing this lab activity, you should be able to:

- Describe the traits of zones in lakes and ponds, including the littoral zone, the profundal zone, and the pelagial zone.
- Relate the amount of carbon dioxide (CO_2) dissolved in the water and water pH.
- Talk about the relationship between water temperature, the amount of oxygen (O_2) dissolved in the water, and the activities of aquatic plants and animals.
- Relate the depth of light penetration in a lake to the amount of photosynthesis by plants and algae.
- Make and interpret measurements of the physical parameters of a body of water including surface water temperature and light penetration in two locations.
- Measure and interpret several chemical parameters of a body of water, including pH and dissolved oxygen.
- Determine whether a pond, lake, or stream would be good for commercial fish production based on the water chemistry measurements you made.

CAUTION:
Several chemicals that could cause burns or skin irritation are used in water testing. Safety goggles with splash guards should be worn at all times when handling these chemicals. Check with your teacher about the correct way to dispose of these chemicals when you have finished. DO NOT pour test reagents down the drain without asking your teacher first. When handling water samples, it is always a good idea to wear gloves and to wash your hands thoroughly with antibacterial soap when you are finished.

Recommended Resources:
Read Chapter 17 in *Exploring Agriscience, 5th Edition.*

Introduction

Commercial fish growers need to understand the factors that affect the water quality in their ponds, lakes, and streams. Producing good-tasting fish with a high yield is possible only when the water quality is kept within an optimal range for the type of fish grown. All commercial fish producers must be able to measure water quality in their ponds. This laboratory exercise focuses on the physical and chemical traits of freshwater ponds and lakes.

Water Temperature and Light Penetration

The structure of lakes and ponds is related to the water level and the pattern of plant life in the lake. Plants and algae are important sources of food for fish and assist in keeping water **pH** and dissolved oxygen levels in lakes and ponds within the range preferred by fish. Plants and algae remove carbon dioxide (CO_2) and add oxygen (O_2) during photosynthesis. A pond or lake can be divided into three distinct regions. The **littoral zone** is the part of the shore and lake bottom between the highest water level and the part of the lake bottom where submerged rooted plants grow. The bottom of the lake, the **profundal zone**, contains sediments from which plants and algae are absent. This is where sediment collects. The **pelagial zone** is the part of the lake or pond that is too deep for rooted plants to grow—open water. Algae and fish are common in the pelagial zone of a lake.

Water temperature affects the water quality and the organisms living in the pond. Oxygen is needed to support the metabolic activities of aquatic organisms. As water temperature increases, the ability of water to hold oxygen decreases (see Table 17-1). But as water temperature increases, the fish in a pond or lake also need more oxygen. Oxygen demand is highest in commercial fish ponds in the summer, when dissolved oxygen concentrations are lower. During warm weather, commercial fish producers may have to add oxygen to the water by mechanically stirring the water—aeration. The water in shallow ponds in which light penetrates to the bottom is uniformly warm.

Activity: Measuring Physical Water Parameters

Materials Needed

- A mercury or alcohol thermometer
- String (to tie onto the thermometer)
- A Secchi disk

In this part of the laboratory exercise, you make measurements on two different ponds or lakes chosen by your teacher. One of the ponds should have some source of nutrients such as fertilizer, pasture runoff, or fish food. The other pond or lake should be a natural body of water that is not actively managed for fish production. Some of these measurements can be made in your laboratory

Water Temperature (°C)	Oxygen (mg/l)
0	14.621
1	14.216
2	13.829
3	13.460
4	13.107
5	12.770
6	12.447
7	12.139
8	11.843
9	11.559
10	11.228
11	11.027
12	10.777
13	10.537
14	10.306
15	10.084
16	9.870
17	9.665
18	9.467
19	9.276
20	9.092
21	8.915
22	8.743
23	8.578
24	8.418
25	8.263
26	8.113
27	7.968
28	7.827
29	7.691
30	7.558
31	7.430
32	7.305
33	7.183
34	7.065
35	6.949
36	6.837
37	6.727
38	6.620
39	6.515
40	6.412

Table 17-1. Oxygen solubility as a function of water temperature.

classroom on water samples collected at the two sites. Other measurements must be made at the pond and will require a field trip. Remember to follow carefully the directions given to you by your teacher to prevent accidental injury. You will make several observations on the two bodies of water: surface water temperature and light penetration. Record the results of your observations in Table 17-2.

Procedure

1. Description of Collection Sites: You need to record important information that describes the two ponds or lakes chosen by your instructor. Include information about the size of the pond or lake, how deep it is (if known), the kinds of plants surrounding the water, and any information about land use in the areas next to the pond or lake.
2. Temperature: A mercury or alcohol thermometer can be used to make surface water temperature readings. Tie a string to one end of the thermometer and dip the thermometer just below the water's surface in a sunny area. Wait three to five minutes before removing the thermometer from the water and reading the water temperature (°C). Record the temperature in Table 17-2. Repeat these steps in a shaded area and record the temperature value (°C) in Table 17-2.
3. Light Penetration: A Secchi disk (see Figure 17-1) can be used to determine the depth to which light penetrates in a pond or lake. The Secchi disk is a weighted 20 cm diameter disk with white and black alternating areas. The Secchi disk is attached at the center of the disk to a line with calibrations at regular intervals to measure depth (usually in meters). The same person should make two observations at each pond and take the average of the two readings. Record your results in Table 17-2.
4. Upon completion consider the following:
   a. Which pond (or lake) had the highest water temperature?

   b. Which pond (or lake) had the best light penetration?

c. Which of the two pond sites has more living plants and animals?

Physical Parameter	POND SITE 1 (with chemical input)		POND SITE 2 (Natural)	
Surface water temperature (°C)	Sunny Area	Shaded Area	Sunny Area	Shaded Area
	(A)	(B)	(A)	(B)
	Average (A + B)/2 =		Average (A + B)/2 =	
Light penetration (meters)	(A)	(B)	(A)	(B)
	Average (A + B)/2 =		Average (A + B)/2 =	

Table 17-2. Measurement of physical parameters for two different pond sites.

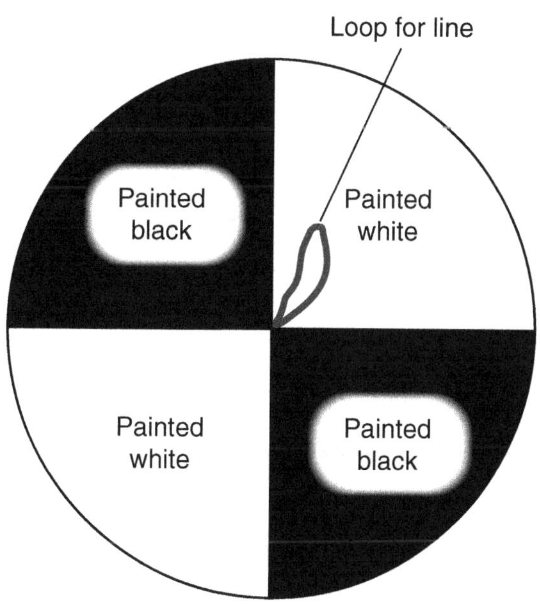

Figure 17-1. Secchi disk.

Questions for Thought

1. Did light penetrate better in the sunny areas or the shaded areas? Explain your answer.

2. Which of the two pond sites is better suited for fish production? Why?

3. Label the zones of a pond, as shown in Figure 17-2 as best you can.

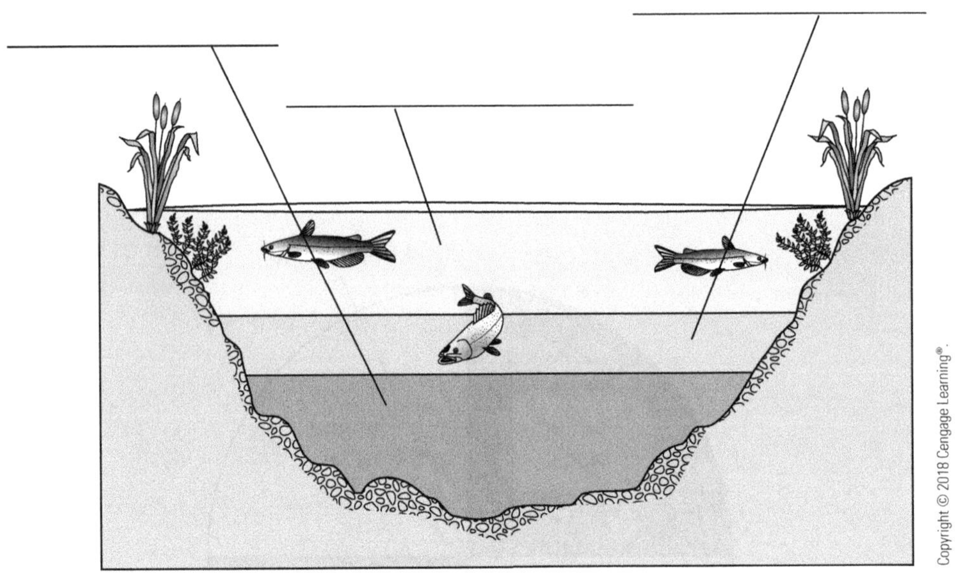

Figure 17-2. Zones of a pond. Label the zones on the figure: littoral zone, profundal zone, and pelagial zone.

Chemical Properties of Water

Dissolved oxygen (DO) is required by aquatic organisms. Accurate and frequent DO measurements should be made in all aquaculture ponds to be sure that there is enough oxygen for the fish. Dissolved oxygen values of 5.0 mg/l are required for fish to grow (see Figure 17-3). Dissolved oxygen values change with the seasons and

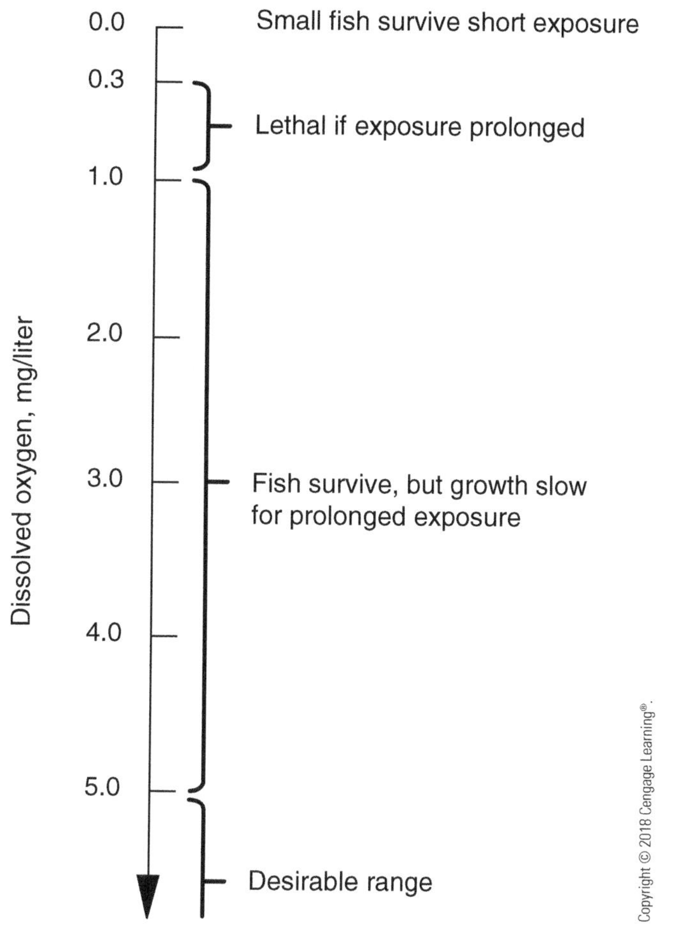

Figure 17-3. Dissolved oxygen and the growth of warm-water fish.

fluctuate daily. All things being equal, dissolved oxygen increases as water temperature decreases. Cold-water fish and fish found in streams with fast currents (such as speckled trout) have high dissolved oxygen requirements. Warm-water fish such as Tilapia demand less dissolved oxygen than their cold-water counterparts.

Dissolved oxygen levels are usually higher near plants, and DO values peak between noon and 6:00 p.m. as a result of photosynthetic activity.

Carbon dioxide (CO_2) is produced by respiring plants and animals. Carbon dioxide dissolves in water. Carbon dioxide dissolved in water (H_2O) forms carbonic acid. The water pH of a pond, lake, or stream is related to the amount of carbon dioxide in the water. The pH scale ranges from 1 to 14; 1 is a strong acid and 14 is a strong base (alkaline). Neutral pH is 7.0. Soils derived from limestone tend to produce ponds with slightly alkaline water. These ponds have a greater capacity for absorbing any excess carbon dioxide produced by fish. Many commercial catfish ponds are located in the southeastern and Midwestern states in areas with alkaline limestone soils.

Water within the range of pH 6.5 to 9.0 is best for raising fish (see Figure 17-4). Extra carbon dioxide produced by fish as a result of overcrowding can reduce water pH to as low as 4.5. Large numbers of bacteria growing

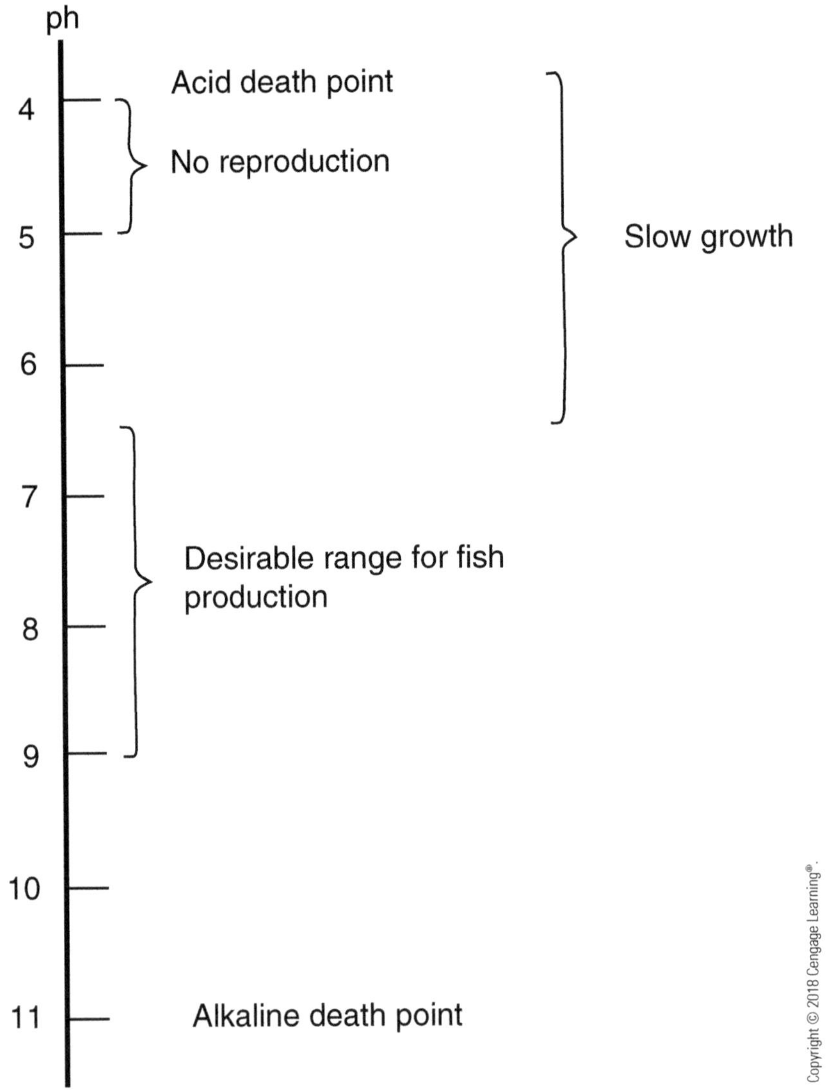

Figure 17-4. Effects of pH on warm-water fish production.

on fish food or chemical fertilizers can cause a drop in water pH. A sudden increase in the amount of algae in a pond or lake can cause a severe drop in water pH. The algae use most of the oxygen for photosynthesis, and an absence of light beneath the algal mat prevents the oxygen in the air from dissolving in the water. As the fish struggle to get oxygen, they tend to use any dissolved oxygen and also add dissolved carbon dioxide to the water, lowering the pH of the water. It is not unusual to see fish die during large algal blooms. The water chemistry of a pond or lake shifts with day/night cycles. The water in a pond is most acidic when it is dark. The pH rises as the day progresses. Water pH is greatest at about 6:00 P.M. (see Figure 17-5).

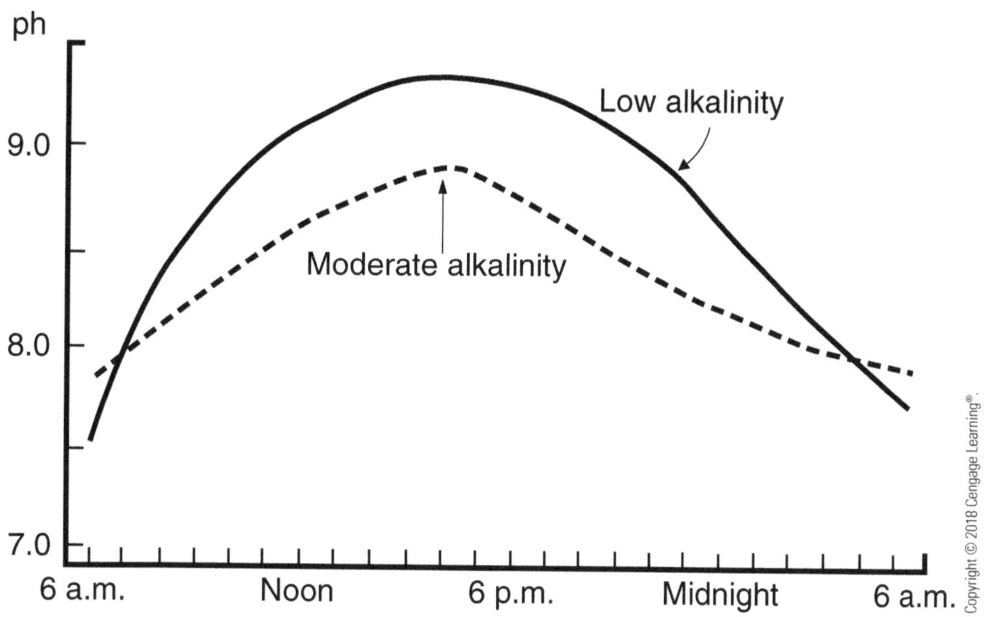

Figure 17-5. Changes in water pH during a day.

Activity: Measuring Chemical Water Parameters

You will estimate several water chemistry parameters using field test kits such as a Hach® or LaMotte® kit on two different bodies of water chosen by your teacher. One of the ponds should have some additional source of nutrients such as fertilizer, pasture runoff, or fish food. The other water source should be a natural body of water that is not actively managed for fish production.

- **pH**-As mentioned, pH affects many water chemistry parameters. Measurements of water pH should be made at the pond or lake. The most accurate pH measurements are made with a field pH probe; pH indicator paper can be used, but it is not extremely accurate. If you are using a Hach® or LaMotte® water chemistry kit, colorimetric methods (matching a colored solution) for estimating water pH are included in the test kits.

- **Dissolved oxygen** can be measured with a probe or estimated using the titration methods found in water test kits.

CAUTION:
The reagents used in estimating dissolved oxygen by titration are caustic and toxic. Follow the directions in the test kit carefully. When you are finished, follow the directions given by your teacher to dispose of the sample.

Materials Needed

- A field pH probe or pH indicator paper
- Dissolved oxygen meter
- BOD (biological oxygen demand) bottle
- A magnetic stirrer
- Two magnetic stirrer bars
- Distilled water

Procedure

1. Using pH paper or a water test kit, Measure the dissolved oxygen in the area with rooted plants and in the open water of the two pond sites and record the values in Table 17-3.
2. Using pH paper or a water test kit, measure the pH of the water in an area of the pond near the shore where plants are present and in the area of open water. Record your values in Table 17-3.
3. Rinse the probe with distilled water and place it in a BOD bottle filled with distilled water. Place a magnetic stirrer bar in the BOD bottle. Turn on the meter and allow it to come to equilibrium.
4. Use a BOD bottle to take a sample. Place a magnetic stirrer bar in the BOD bottle. Place the BOD bottle on a magnetic stirrer and turn the stirrer on, slowly increasing the speed. Take the measurement, allowing one to two minutes for the reading to stabilize. Record the value in Table 17-3.
5. Rinse the probe with distilled water, return it to the BOD bottle filled with distilled water, and allow the probe to come to equilibrium. Repeat the process for the remaining water samples.
6. When you are finished, rinse the probe with distilled water and return it to the BOD bottle containing distilled water. Turn the meter off, but do not allow the probe to dry out.

Chemical Parameter	POND SITE 1 (with nutrients added)		POND SITE 2 (Natural)	
	Plants	Open Water	Plants	Open Water
pH				
Dissolved Oxygen (mg/l)				

Table 17-3. Water chemistry parameters for littoral and pelagial zones of two ponds.

Questions for Thought

1. Were your results what you expected? Which pond or lake had the highest dissolved oxygen concentrations?

2. Were there differences between the water surrounding the rooted plants and water in the open part of each pond or lake?

3. Would you expect more fish in the open area or around the roots of plants? Support your answer with evidence from Table 17-3.

4. Which site is best for plant growth (including algae)?

5. Which site would be the best location for a commercial fishery? Supply evidence for your decision based on the results you collected in Tables 17-2 and 17-3.

Glossary

littoral zone: the zone of a lake, pond, or ocean between the highest seasonal **water** level and the area inhabited by submerged rooted aquatic plants

pelagial zone: the zone of open water in a pond, lake, or ocean

pH: a measure of the relative concentration of hydronium (H^+) ions. The greater the H^+ concentration, the more acidic a substance is; the lower the concentration of H^+ ions, the more alkaline a substance is.

1		7		14
very acidic	acidic	neutral	alkaline	very alkaline

profundal zone: the portion of the bottom of a body of water that is devoid of vegetation or microorganisms; characterized by deposits of sediments.

EXERCISE

18

Companion Animals

Student Objectives

After completing this lab activity, you should be able to:

- Describe the basic steps in the problem-solving method.
- Identify advantages and disadvantages for animal species based on different living situations.
- Use problem solving to determine the best animal species for a specific person or family situation.

Recommended Resources:

Read Chapter 19 *in Exploring Agriscience, 5th Edition.*

Introduction

Companion animals are not limited to dogs and cats. As discussed in the text, companion animals include just about all animals. While all animals have the potential to be companion animals, not every setting is capable of handling all animals. For example, you would not be able to house a full-sized horse in an apartment in New York City. As much as a person may want a horse, the animal still has personal needs that must be met, and an apartment does not meet those needs.

Some creative people came up with an alternative to a full-sized horse living in their house. They discovered that miniature horses have basic need requirements much like dogs. The size of a miniature horse, food requirements, and daily waste generated were very similar to those of a large dog. As a result, they could now keep a horse in their house without dealing with all the basic requirements of a full-sized horse. These people used problem solving to help find an answer to their situation.

Problem Solving

Problem solving is a method of dealing with a specific issue, by working through several steps to reach the best answer for the problem. The first step is to understand the specific issue that needs to be solved. If you do not understand the specific issue you will never be able to determine the central problem.

Once you understand the specific issue, you need to identify all the potential problems with the specific issue. There can be as few as one potential problem or more than twenty. In raising a horse indoors, there are several potential problems: large daily feed and water requirements, animal waste removal from household surfaces, and a large animal breaking furniture and bumping into walls. Once the group brainstorms all the potential problems, they should be able to review them and come up with a central problem.

Often there are many different problems that make up the central problem. All the potential problems should point to a central problem. In the case of raising a horse indoors, the central problem is "How can we keep a horse in our house without all the housekeeping issues associated with raising a full-sized horse indoors?"

Once you define a central problem, you can brainstorm a series of solutions that can solve the problem. Solutions could be as straightforward as building a bigger house that is horse friendly. Another solution could be as off-the-wall as moving into a barn and living there, making it your new house. Any idea has the potential to solve the problem, no matter how crazy it may first seem. Once you have come up with a series of solutions, the group needs to choose the ultimate solution to solve the problem. In the case of the people wanting to raise the horse, the group decided to try and find a miniature horse. That way, they have all the fun of the horse without turning their house into a barn.

Activity: Companion Animals and Problem Solving

Given here are several situations. As a lab group, you must use the problem-solving method to find the ultimate solution for the companion animal's needs. Please review the problem-solving steps:

- Identify the specific issue.
- Brainstorm potential problems with the specific issue.
- List the central problem.
- Brainstorm solutions to the central problem.
- Create the ultimate solution to the central problem.

Situation #1

A young couple has two small children (one infant, one toddler). The parents feel that the two children need a pet to learn the basics of care for another animal.

However, in the case of the infant, they are concerned about potential allergic reactions caused by some animals. In the case of the toddler, the parents are concerned about the aggressiveness of some animal species because the child is able to move around by herself. The parents understand that they will largely be responsible for the day-to-day care of the animal, but they would like to keep that to a minimum. What animal would be best for this young family?

Situation #2

This situation is similar to #1 except that the children are older and the parents want the children to take full responsibility for the animal. The parents want an animal that can interact with the family but is not so large that it dominates the children. The parents want minimal responsibility, so animals that will shed and leave hair all over the house are not an option. Animals that leave strong odors in the house are not welcome either. What would be the best animal for this family?

Situation #3

A retired couple wants an animal that is like another grandchild. They want to take this animal everywhere with them, and they wish to give it the best care possible. They want an animal with a lot of personality and that is trainable, so that not only can it be housebroken but it can go outside and respond to commands. It must interact well with the couple's grandchildren and have a long lifespan so that they can enjoy it for many years. What animal would be best for this family?

Situation #4

A single male wants an animal. He is a professional who is away from home several days a week, but he has someone who can feed it when he is away. His comment is, "I basically want an animal that does not require much more attention than a plant." He would like something exotic and that few people have as a pet. He would prefer something with lots of colors or style that people would want to talk about it when he has parties or meetings at home. What animal would be best for this guy?

Questions for Thought

1. What part of the problem-solving process was the most difficult to complete? Explain why.

2. Based on the above examples, do you now see a type of companion animal or category of animals that really are not good pets for most people? List them and give reasons why.

3. What limiting factor came up most often in trying to choose and animal? For example, purchase price of the animal and related equipment, healthcare requirements, daily attention requirements for the animal, daily housekeeping requirements for the animal, and others. Explain why the factor was such a problem.

EXERCISE

19

Preserving Food

Student Objectives

After completing this lab activity, you should be able to:

- Describe differences between fresh and preserved products.
- Discuss the advantages and disadvantages to preserving food.
- List types of food that are better suited for long-term preservation.

Recommended Resources:

Read Chapter 20 in *Exploring Agriscience, 5th Edition.*

Introduction

Before the age of mechanical refrigeration, the notion of storing food was held to things that would not spoil easily or quickly (e.g., nuts, grains). People were forced to eat fruits and vegetables as they became and remained ripe. Storing these items as we do today (e.g., by refrigeration, freeze-drying, canning, jarring, or freezing) was not an option. Communities or families could not efficiently keep fresh food for weeks or months in a safe state. Instead, they were forced to harvest and eat only what was necessary. Over time, and facing many challenges with food supply, our society adapted, and the result was the development of various methods of food preservation.

Drying Food

One time-tested method for food preservation is drying, and is likely the first method of food preservation developed. In the beginning, people used the sun and fire to dry foods. Even without refrigeration, people were able to dry fruits, meats, and vegetables, into products like dried apricots, beef jerky, or dried beans. This method worked so well, it has survived to present day. Today, this task can be accomplished through the use of a food dehydrator, as demonstrated in this lab exercise.

Activity: Dehydration of Food

In this activity you have an opportunity to taste both raw and dried fruits and vegetables, allowing you to distinguish flavors, textures, and consistency of foods in both an old **preserved** state and a fresh state.

CAUTION:
If you have any food allergies, please inform your teacher. Work with a partner who can make observations on taste and communicate them to you.

Materials Needed

- A produce item (fruit or vegetable)—one per student
- A paring (or other slicing) knife
- Cutting board
- Sink
- Food dehydrating machine (in the lab)

Procedure

1. Bring in a produce item from home or the grocery store.
2. Wash your item's exterior to remove any dirt, insects, or chemicals that may have been applied prior to purchase.
3. In the lab, using a paring knife, slice or cut your item on a cutting board or surface into bite-size chunks or pieces to share with your classmates. Be sure to cut your item into enough pieces that each of your classmates can have a sample in both raw and **dehydrated** form.
4. Take a small portion of your produce item and place it in the food dehydrator.
5. While your item is dehydrating, rotate with your classmates, tasting each item around the class and take notes regarding flavor, consistency, and texture. In a chart, note the flavor (or lack thereof) in each item's raw form.
6. Before class is complete, make sure to carefully wash your cutting surface and knife. Be extra careful handling sharp items, especially when carrying them to or from the sink.

7. Take a sample of at least five dehydrated items and again make note of flavor, texture, and consistency.
   a. Describe below how the dried versus raw forms of each item differ.

   b. Which items tasted better or worse when compared to the original raw form?

Variation

Should a food dehydrator not be available, the class can be divided in half, with half of the students bringing fresh produce and the other half bringing in dried produce (available in your local grocery store or food co-op).

Questions for Thought

1. What are the advantages and disadvantages to preserving various food sources?

2. How has the ability to preserve food helped to move our society forward? What are we now able to accomplish with food preservation?

Glossary

dehydration: the process from removing water from a substance or compound

preservation: to prepare a food so as to resist decomposition or fermentation

EXERCISE 20

The Ethical Treatment of Animals

Student Objectives

After completing this lab activity, you should be able to:

- Describe the advocacy groups associated with the ethical treatment of animals.
- Research animal rights issues and understand the reliability of the information they obtain during research.
- Provide facts and constructive arguments for and against various animal confinement methods used in agriculture.

Recommended Resources:

Read Chapter 21 in *Exploring Agriscience, 5th Edition.*

Introduction

Animal welfare remains a very controversial topic among various groups. Opinions on what is proper treatment vary based on the background of the people associated with these groups. Information and understanding of agricultural practices often help students better understand why animal confinement practices like farrowing crates, feedlots, and broiler cages are used in the related animal production industries. However, understanding the use of this equipment often does not end the debate.

Fact-Based Discussion

Animal rights advocates believe that animals should be free and have the same rights to live as do humans. **A animal welfare advocates** stress the importance of letting animals live as they would in nature, but agriculture industry supporters counter that more natural methods used in production increase the cost of farming and that these practices are not as healthy for the animals as they may appear.

A fact-based discussion can provide a good opportunity for all sides to be heard on any topic. The key to finding facts to support your opinion is to research the subject and use references from reliable sources. While the Internet does provide a large amount of information, much of that information is not peer reviewed. The danger of using information that was not reviewed is that the information could have simply been made up by the author or could be merely the author's opinion.

Peer-reviewed articles are published in scientific journals and are reviewed by others in the industry before they are allowed to be presented to the general public. This extra review step allows other individuals in that scientific discipline to provide their knowledge on the topic and ensure all the information is factual and supporting of the topic discussed. These peer-reviewed articles provide a much more trustworthy source of facts to support your opinion on a topic.

Activity: A Debate on Animal Confinement in Agriculture

Below is a list of animals and confinement issues that can be debated:

Swine—**farrowing crates** are used to limit sow movement, and therefore reduce piglet injury during the birthing process

Poultry—broiler crates are used to prevent escape and injury of the birds during the transportation of chickens from the broiler house to the processing plant

Poultry—layer cages are used to limit movement and injury of hens in a layer house while they produce eggs

Cows—**farm factories** (feedlots) are used to house large volumes of cattle and control their feed intake while increasing their weight before they are taken to a processing facility

1. Break up into your lab group and review the advocacy groups listed:

   Animal rights group—This group focuses on keeping all animals as free as possible and free to live as humans do.

   Animal welfare group—This group focuses on allowing humans to use animals while providing for their comfort and proper care.

   Animal industry—This group, while similar to the animal welfare group, sees animals as a resource for humans. Many believe proper care makes for more productive animals. They also see these animals as a product for a business and acknowledge that cost is still a factor in raising animals.

   General public—This group has opinions that will vary widely from person to person, but there are some key points they focus on. The general public is concerned about the welfare of animals and are not in favor of animal abuse. They are also concerned about the cost of food they buy.

2. Research the confinement issue selected by the teacher from the perspective of the advocacy group assigned to your lab group. Remember, in gen-

eral, the research sources found either in the library or on the Internet may contain information that is not always reliable. Review the list of information sources below, which are in order of reliability/truthfulness. The sources at the top of the list are the most reliable and the ones at the bottom are the least reliable.

- Peer-reviewed textbooks, books, and scientific journals (many of these are available online through various universities and scientific organizations)
- Magazine articles
- Newspaper articles
- Videos and television stories
- Internet articles (with a listed company, university, or association) with citations
- Internet articles (blogs, webpages) without citations

3. Prepare your arguments and facts in a manner that will be easily readable during the classroom discussion on the topic. The teacher should provide you with the format for the discussion to make preparation easier for you.

Questions for Thought

1. Is animal confinement, while not part of the natural life of the animal, better for the animal's well-being or worse?

2. Is there a more acceptable alternative to the confinement method that is just as effective? If so, is it used currently? If not, why not?

3. If the confinement method was banned, would you be satisfied with the market price of the meat, eggs, leather, and other goods? What if the price doubled? Or tripled?

4. Create a law that would be beneficial to the animal being confined with minimal impact on the producer. What would be the drawbacks of your law?

Glossary

animal welfare advocates: people who feel that it is all right to raise animals for human use but that they should be cared for properly and the animals should be comfortable in their surroundings.

animal rights advocates: people who feel that animals should be free and that they have as much right to live as do humans.

farm factories: confinement operations that produce the maximum amount of animal product in a minimum amount of space.

farrowing crates: a confinement device used by swine producers during the birthing of piglets.

EXERCISE

21

Using Hand Tools

Student Objectives

After completing this lab activity, you should be able to:

- Practice safe use of some hand tools.
- Accurately demonstrate use of a combination square.
- Describe when it is appropriate to use a chalk line versus a pencil.
- Describe how to properly clean hand tools.

Recommended Resources:

Read Chapters 1 and 22 in *Exploring Agriscience, 5th Edition.*

Introduction

Being able to apply measuring skills, learned earlier, is an essential skill for anyone using hand or power tools. Whether you are repairing damaged materials or constructing new projects, it is critical for you to know how to measure accurately and use the proper tools to do so. With the passage of time, you will find numerous projects in which you will be handling wood. This is one activity in which you learn things that lend themselves to many other activities and projects.

Proper Use of Hand Tools

In this exercise you learn to square off boards, using a combination square to ensure that they are squared off at the edges, and you also learn to correctly apply a brace and utilize an auger bit to drill holes in these boards. A **combination square** typically has a steel or metal straight edge accompanied by a scriber that is housed in the head, along with a built-in level.

To ensure that you understand the use and application of squares or other measuring tools, you and/or your teacher may discuss various purposes for having boards squared off.

NOTE: *You are responsible for cleaning up your work area and any tools used in this activity.*

Activity: Using Tools Properly

Materials Needed

- Brace or clamps
- Auger bit
- Drill
- Standard ruler
- Combination square
- Chalk line
- Pencil
- Workbench
- Brace or clamps
- 2 × 8 boards in 3-foot lengths for each participant or small group depending on resources available
- Safety goggles for each participant
- Brush or handheld broom for cleanup

Procedure

NOTE: *You may need to share these tools (such as a chalk line or other pieces) with other students in the class. You must begin by wearing appropriate safety clothing (e.g., safety glasses).*

1. Take a wooden board and lay it flat on your work surface.
2. Using the ruler, find a point approximately 3 inches from one end of the wooden board.
3. Using the combination square and a pencil, mark a straight line from the top edge to the bottom edge of the board at the 3-inch mark that you made in Step 2. This line should be perpendicular to the long edge of the board. This is your squared edge line.
4. Repeat Steps 2 and 3 at the other end of the wooden board. However, this time instead of using a pencil to mark the squared (perpendicular) edge, use a chalk line; this gives you a chance to use a different tool for a similar purpose. Snap the line at the edge of the combination square to mark the straight edge or squared edge.
5. Using a brace or clamp, secure your board to your work surface in preparation for drilling. Secure your board so that it lies flat as in Step 1.

6. Looking at the pencil line drawn in Step 3, use your ruler to measure 2 inches toward the center of the board and 2 inches from the top edge of the board. Mark the point where these two measurements intersect with a pencil mark or an X. See the accompanying illustration (not to scale).

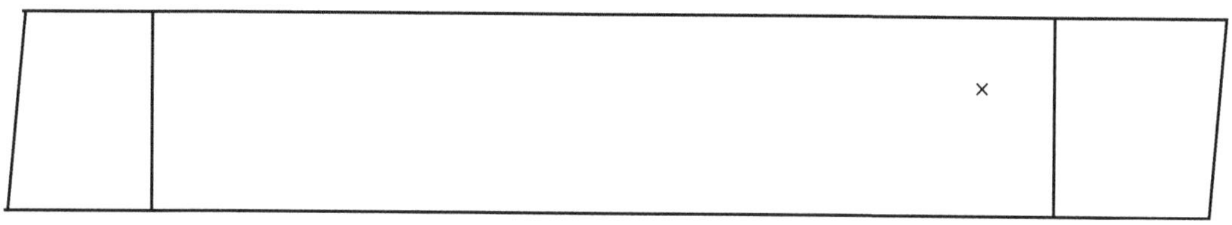

7. Similar to Step 6, measure 2 inches from the squared edge line and 2 inches from the bottom of the board, and again mark with a pencil mark or an X.
8. Repeat Steps 6 and 7 at the other end of the board.
9. You should now have four pencil marks in similar locations on the board.
10. Remove the clamps on your board, and slide your board so that one of your marks is sitting over the edge of the work surface (so that you do not drill through the work surface when you drill). Secure your board with the brace or clamps.
11. With the power off and the safety engaged, take your drill and insert the auger bit securely into the drill.
12. Carefully place the drill bit perpendicular to the board at your mark, carefully engage the drill, and press the drill bit through the board. Keep the drill bit engaged while pulling the drill bit back through the newly drilled hole, still perpendicular to the board.
13. Repeat Steps 10 through 12 until all four points have been drilled and your board has four holes instead of four marks.
14. Turn the drill power off and unplug it (if necessary). Secure the safety lock and remove the auger bit from the drill.
15. Using a brush or other cleaning tool, remove any remaining wood shavings from the bit and place it back in its original case or location.
16. Using a dry cloth or rag, wipe off the exterior of the drill, your work surface, and any tools used to remove any loose dust or wood shavings. Place the tools in their original location.
17. Using a dust pan and broom, sweep up any remaining dirt or dust in your work area created by this activity.

Questions for Thought

1. What is the benefit of using a combination square compared to just a ruler?

2. Which method of marking was easier? Why? When might a chalk line be used instead of a pencil?

3. Why is it important to clamp your material when drilling or cutting?

Glossary

combination square: a squaring tool with a movable head used to check whether a board is square at the end, or to lay out square lines. A combination square is more functional as it can do the equivalent of try squares or framing squares, in addition to being used for 45° or 90° angles.

EXERCISE

22

Small Engines

Student Objectives

After completing this lab activity, you should be able to:

- Describe the servicing process for both two-cycle and four-cycle small engines.
- Diagnose common problems with both two-cycle and four-cycle small engines.
- Describe the importance of regular service and maintenance of small engines.

Recommended Resources:

Read Chapter 23 in *Exploring Agriscience, 5th Edition.*

Introduction

Small engines impact our daily lives whether we know it or not. They not only cut grass and prune shrubbery but also power portable generators and water pumps. The latter two are quite useful during power outages and flooding. Many homeowners put off regular maintenance of their small engines, but in fact most people can take the time to make repairs and still manage to complete the grass cutting or trimming the same day. However, if there is a power outage or a flood event, the homeowner needs the small engine to start up the first time. That is why regular maintenance is so important for every small engine you own.

Servicing Small Engines

Service of a small engine should occur as regularly as you cut the grass. Service should become a habit that in time will save you far more time and lead to a better performing engine. However, there are occasions where regular maintenance and service of the small engine is not enough and your small engine will not operate correctly. That is when you need to learn the steps of troubleshooting. **Troubleshooting** is the act of diagnosing what is wrong with a small engine based on what the small engine will or will not do for the owner.

Two-Cycle Engines

One benefit of two-cycle engines is that regular service does not include checking the oil level. These small engines use fuel that is a combination of gasoline and oil to operate. However, before every start of a two-cycle engine, the fuel level should be checked. While this may seem obvious, it is important to remove the gasoline cap and look into the fuel reservoir to make sure there is no floating debris or water contained in the fuel.

Also, if the small engine has been in storage for a long period of time, checking the fuel level can indicate whether the fuel may need to be disposed of and new mixed gas added. Often fuel that remains in a container for an extended period of time will spoil and cause the engine to operate improperly—if it operates at all. One indication of this is that mixed gas gives off an odor that is even less appealing than fresh mixed fuel.

Pull-start small engines should have their pull handle and string inspected before every use. Fraying of the rope or a cracked handle indicates a need to replace these parts before they break during engine startup. Breakage during startup can cause serious injury to the operator.

Finally, two-cycle engines are air cooled, and the cooling fins on the outside of the engine should be inspected before each use for dirt, mud, or other debris that could keep the engine from cooling properly. An overheated engine can affect operation and performance during use of the equipment.

Generally, after several uses (around 10 hours) you should also inspect the oil filter of the small engine. This filter is often a synthetic foam screen over the air intake. Grass and dirt as well as any other debris should be removed from the filter. Once cleaned, the filter should be reoiled and mounted back onto the small engine.

Several times a season (every 25 hours), the spark plug condition and gap should be inspected. The **gap** is the distance from the center electrode to the side electrode. The **side electrode** is the small metal arm that arcs above the center electrode. Buildup of soot, oil, and carbon are indications that there are other problems with the small engine. The electrodes should have their factory edges and even thickness. Thinned metal or rounded edges also indicate that the small engine is not operating properly.

Four-Cycle Engines

Regular service of four-cycle engines will not vary drastically from that of two-cycle engines. The oil level should be checked before operating the small engine. Additional oil should be added based on the manufacturer's specifications. The fuel level, pull start, and cooling fins should be inspected as well.

One difference between four-cycle and two-cycle engines is the air cleaner element on the four-cycle engine, but this should be inspected similarly to the two-cycle engine (every 10 hours). The air cleaner element often consists of two parts: a pre-cleaner and a paper element. This air filter is often referred to as a paper element because of the material it is made of. Inspect the paper element for dust, grass, and other debris. A severely clogged filter should be replaced. The pre-cleaner is often made of small-cell foam and should also be inspected for debris. If it is dirty, remove the pre-cleaner and wash it with warm water and liquid detergent. Never oil this pre-cleaner as you do on a two-cycle engine. Finally, inspect the air cleaner element housing for any debris that could be drawn into the engine, and remove it.

As with a two-cycle engine, the spark plug condition should be checked (every 25 hours). Diagnosing problems with the spark plug on a four-cycle engine will be similar to the process for a two-cycle engine.

Activity: Diagnosing Small Engine Problems

Materials Needed

- Several small engines
- Index cards with particular problems described
- Notebook paper and pen or pencil

Procedure

1. On a clean sheet of paper, list the station number and situation stated on the index card at your small engine station.
2. List the steps you took to diagnose the problem. For example, "inspected the spark plug," "checked oil level."
3. Indicate what your final diagnosis is for that station and suggest a repair.
4. Spend 5 to 10 minutes at each station and complete the diagnosis before moving to the next station.

Troubleshooting Hints

- Poor performance—contaminated fuel, spark plug not operating effectively, air intake clogged
- Engine will not start—no fuel, clogged fuel line, contaminated fuel, spark plug faulty, spark plug wire unplugged or damaged
- Engine overheating—air intake screen clogged, cooling fins plugged, not the proper amount of engine oil
- Engine will not idle—spark plug not gapped correctly, faulty spark plug

Questions for Thought

1. What is the value of regular maintenance of small engines?

2. List some safety considerations that must be addressed when servicing or repairing a small engine.

3. List several problems that cause small engines not to work. Then describe what regular maintenance procedures would have located the problem beforehand and prevented damage or delay of use to the small engine.

Glossary

gap: the distance from the center electrode to the side electrode

side electrode: the small metal arm that arcs above the center electrode

troubleshooting: the act of diagnosing what is wrong with a small engine based on what the small engine will or will not do for the owner

EXERCISE

23

Biofuels

Student Objectives

After completing this lab activity, you should be able to:

- Describe the basic soybean oil extraction process.
- Calculate the amount of oil needed to produce one gallon of biodiesel.
- Name the three alternative fuel sources created from crops.

Recommended Resources:

Read Chapter 24 in *Exploring Agriscience, 5th Edition.*

Introduction

Energy companies are looking to various plant species to provide the needed materials to refine into biofuels. Soybeans, switchgrass, and corn are some of the agricultural crops that are currently in production for use as **biofuel**. The benefit of these crops is that they provide a renewable source of fuel. **Ethanol**, **biodiesel**, and **biomass** are the three alternative fuel sources into which crops are categorized. This division helps define the process used to make the biofuel.

Creating Biodiesel from Soybeans

Plant species that contain large quantities of vegetable oil can be refined for use in biodiesel. One of the most commonly used agricultural crops for biodiesel production is the soybean. The large quantity of vegetable oil that is extracted from the soybean, compared to other agricultural crops, makes it the best choice for biodiesel production. Yield of oil per soybean is one of the most critical components of biodiesel creation. Various techniques to crush and flatten each bean are used to quickly remove the oil and process the beans. One of the most common methods involves first cracking each bean and removing the thick outer hull. The remaining parts are further processed by a method called flaking. The beans are moved through large rollers that press each bean into "flakes." Once all the material is flaked it is moved to large holding tanks where it is soaked in a special chemical solution that removes 99 percent of the available soybean oil from the beans.

Activity: Creating Biodiesel from Soybeans

Materials Needed

- Fifty soybeans
- Two paper plates: one for the raw beans and one for the crushed beans
- A graduated cylinder
- A funnel
- Two metal or sturdy plastic teaspoons
- Optional: adjustable pliers or a C-clamp can also provide the needed force to press the bean flat and remove the oil

Procedure

1. Get a minimum of 50 soybeans from your teacher: you will crush and extract oil from these.
2. Rest the funnel on top of the graduated cylinder. Each student in the lab group should take turns attempting the crushing process. You will each place a bean into the crushing apparatus.
3. Once your materials are positioned correctly, hold the bean and crusher over the funnel and apply the needed force to crush the bean as flat as you can to extract the maximum amount of oil from the bean. Take turns to allow your hands to rest between attempts so that you each can provide maximum crush force each time you are required to crush a bean.
4. Open the apparatus and dispose of the bean meal onto the second paper plate. Repeat the process until all the beans are crushed.
5. Detail your results below.

   Number of seeds crushed ____________________

   Milliliters of oil collected ____________________

   Milliliters of oil per seed collected (ml/# of seeds = yield): ______________ ml per seed

Questions for Thought

1. At an average of 180,000 beans per bushel, how many milliliters of oil would you expect to produce through the hand-crushing process? (ml per seed × 180,000) = ________________ ml per bushel.

2. At an average of 40 bushels (bu) per acre, how many milliliters of oil would you yield on one acre of land? (yield per bushel × 40 bu) = ________________ ml per acre.

3. Roughly 7.6 pounds of soybean oil are needed to produce one gallon of biodiesel, and on average 11.28 pounds of soybean oil is produced from a bushel of soybeans. Can you calculate how many gallons of biodiesel can be made through the hand-crushing process?

4. What uses are there for the soybean by-products (i.e., the soybean meal and soybean hulls)?

5. If farmers continue to plant more soybeans on their crop land and sell it to biodiesel refineries, what may happen to the supply of soybeans for other uses?

6. What may happen to the supply of corn and wheat if more cropland is placed into soybean production rather than these other agricultural crops?

Glossary

biofuel: liquid, solid, or gaseous energy sources created from organic materials that come from renewable sources such as agricultural crops

biodiesel: the mixture of plant and animal oils with petroleum-based diesel fuel

biomass: large amounts of organic waste products

ethanol: a form of alcohol produced from the fermentation and distillation of grains

References

Radich, Andy. (2004). *Biodiesel Performance, Cost, and Use.* Energy Information Administration, U.S. Department of Energy.

Westhoff, Pat. (2006). *FAPRI Ethanol Briefing Materials for Congress Peterson—Addendum—Biofuel Conversion Factors.* (FAPRI) Food and Agricultural Policy Research Institute, University of Missouri.

EXERCISE

24

Products of Biotechnology in Agriculture

Student Objectives

After completing this lab activity, you should be able to:

- Describe the effects of *Bacillus thuringiensis* (Bt) toxin on insects through laboratory experiences.
- Determine the importance of disease resistance in plants through laboratory experiences.

Recommended Resources:

Read Chapters 5, 10, 12, 13, and 25 in *Exploring Agriscience, 5th Edition.*

Introduction

Agricultural biotechnology uses biological and **genetic engineering** techniques to change or improve plants, animals, and microorganisms used in agriculture.

An example of a biological technique is the use of a microorganism called *Bacillus thuringiensis* (Bt) to kill many different insect pests that destroy agricultural crops.

Genetic engineering, involves altering the genetic makeup of organisms to produce desired outcomes. Creating a disease-resistance cultivar, or altering microbes to resist frost in order to protect crops, are two such examples of genetic engineering.

You will study all of these examples in this exercise.

Bt Toxins

Bacillus thuringiensis (Bt) is a common soil bacteria bacterium with some very unique abilities.

Diamondback moth	Armyworms
Cotton bollworm	European corn borer
Cabbage looper	Potato tuber moth
Colorado potato beetle	Mosquito
Blackfly	Tobacco hornworm caterpillar

Table 24-1. Insects that are killed by Bt toxin.

It produces toxic proteins called Bt toxins that kill many different larval (young) insects on agricultural crops (see Table 24-1). The Bt toxin is used as a natural insecticide. An **insecticide** is a natural or synthetic chemical that kills insects.

Activity: Study the Effects of Bt as a Natural Insecticide

Materials Needed

- Two to four Tobacco plants (at least six weeks old)
- Twenty hornworm caterpillars
- Bt toxin
- Plastic sheeting
- Stakes
- Tape
- Misting spray bottle
- Tracing paper or plain copy paper

Procedure

1. Label one tobacco plant "Untreated." Place 10 hornworm caterpillars on various leaves of the tobacco plant.
2. Construct a tent over the plant made of plastic sheeting and stakes. Close the tent carefully. Larger classes may need more than one plant.
3. Following the manufacturer's instructions, treat an entire tobacco plant or group of plants with Bt toxin. Label the plant(s) "Bt treated." Place 10 hornworm caterpillars on various different leaves of each plant. Construct another tent, using the steps in step 2, and place it over the plant(s). Close the tent carefully. Larger classes may need more than one plant.
4. It is important to keep the humidity high in the tents. Mist the air in the tent with water at least once a day. Be careful not to wash the Bt treatment off the plants.
5. Over the next week, observe the behavior of the caterpillars. Record any deaths in Table 24-2.

Treatment	Day 1	Day 2	Day 3	Day 4	Day 5
Untreated					
Treated					

Table 24-2. Effectiveness of Bt toxin in controlling tobacco hornworm caterpillar (mortality rate).

6. After one week, remove the leaves of the tobacco plants from all treatments, including the control. Compare leaf damage due to caterpillars between plants.
7. Trace the leaves on paper, including all holes and eaten areas, to serve as a record. Label the traced control and treated leaves.

Questions for Thought

1. Look at the tracings of tobacco leaves from the untreated and Bt-treated plants. Is there a noticeable difference between the areas that are eaten?

2. Was the Bt toxin effective in eliminating the caterpillars?

Protecting Crops with Microbes

Growers around the world lose an estimated $11 billion to $14 billion a year due to frost damage on food crops. Frost damage occurs when ice crystals form on leaves, flowers, and fruits of plants. Water expands as it freezes, killing the cells of the plants and causing loss of flowers and fruits, and even killing whole plants. *Pseudomonas syringae* is a bacterium that grows naturally in soil and on plants. These bacteria make a protein that can help ice crystals in dew or rainwater to form when the temperature is just below freezing. Genetic engineers at a company based in Oakland, California have genetically engineered microbes that do not make the protein that helps ice crystal to form so easily. "Frostban," the genetically engineered "ice-minus" form of *P. syringae*, has been used to prevent frost damage on strawberries and sweet potatoes.

Activity: Protecting Crops with Microbes

Materials Needed

- *P. syringae kit*

Procedure

1. Use a commercially available *P. syringae* kit to observe ice crystal formation in the presence of this ice-nucleating bacteria.
2. Following the direction provided by your teacher, record your observations considering the questions below.

Questions for Thought

1. Describe what you observed.

2. Were the results what you expected? Why/Why not?

Disease-Resistant Crop Plants

The first disease-resistant tomato was discovered in 1912. It was growing in a tomato field that had been badly infested with *fusarium*, a fungus that causes wilt in tomatoes. It was not until the 1940s that the first *fusarium*-resistant tomato cultivar was released for sale. Biotechnology has helped reduce the time it takes to develop disease-resistant cultivars of crop plants such as tomatoes. Most commercially available seeds have been bred to be resistant to a variety of diseases that are caused by bacteria, fungus, and viruses. Some of the more common diseases in tomatoes are *fusarium* wilt, *verticillium* wilt, and tobacco mosaic virus (TMV). Many tomato diseases are present in the soil. The best places to find tomato disease sources are soils that have already been used to grow tomatoes. You may want to bring a bucket full of soil from your home garden for use in testing disease resistance in tomatoes.

Activity: Disease-Resistant Crop Plants

Materials Needed

- Several varieties of nonresistant and disease-resistant tomato seeds
- Small pots
- Soil from a plot that has been used before to grow tomatoes
- Vermiculite
- Labels or tape

Procedure

1. Obtain several varieties of nonresistant and disease-resistant tomato seeds.
2. Fill small pots with soil from a plot that has been used before to grow tomatoes.
3. Plant the tomato seeds in the soil, covering them lightly with vermiculite. Label the pots with your name and the disease-resistant varieties you planted. Water them well and place them in a sunny window or under artificial lights.
4. Allow the tomato seeds to germinate and grow for a period of four to five weeks. Tomatoes are day-neutral plants, so lighting requirements are not strict. A 12-hour light period works well. Evaluate the growth habits of each variety. Some diseases affect only adult plants, so further observations may be necessary.
5. Record your observations in Table 24-3.

Disease Resistance	Week 2	Week 4	Week 6	Week 8	Week 10
Control					
Resistant Variety					
Resistant Variety					

Table 24-3. Disease resistance in tomatoes.

Questions for Thought

1. Were any tomato plants damaged by disease during the trial growths?

2. If there were damaged tomato plants, which ones were they?

3. What observations can you make about the soil that damaged tomatoes were grown in?

Biotechnology and Disease Detection

An important area of research is in testing animals for disease and animal products such as meat and eggs for contamination. *Escherichia coli (E. coli), salmonella*, and *L. monocytogenes* are three food-contaminating bacteria that can cause serious illness in humans. Animals can become very sick after eating feed contaminated with *fusarium*. *Fusarium* fungus produces toxic proteins called **mycotoxins**. Researchers are developing rapid screening tests called ELISAs for use on farms and in food processing plants to detect disease or contamination quickly and easily.

Activity: Disease Detection

Materials Needed

- ELISA testing kit

Procedure

1. Use a commercially available kit to learn how ELISA testing works.
2. Following the direction provided by your teacher, record your observations considering the questions below.

Questions for Thought

1. Describe what you observed.

2. Was the testing process what you expected? Why/Why not?

Glossary

genetic engineering: various techniques used to transfer or change genes in organisms.

insecticide: a natural or synthetic chemical that kills insects.

mycotoxins: toxic substances that come from fungi.

EXERCISE 25

Career Exploration

Student Objectives

After completing this lab activity, you should be able to:

- Identify at least three websites with information for job searching in an Ag-related industry.
- Prepare an outline or overview of basic qualifications for at least three different Ag-related positions.
- List at least three postsecondary schools where you may enroll in a degree program that can prepare you for a job in one of these positions.
- Identify opportunities for employment within the agriculture industry.

Materials Needed:

- Computer, tablet, or smartphone with Internet access; search engine (e.g., Yahoo!, Google)
- Career worksheet (provided)

Recommended Resources:

Read Chapters 26 and 27 in *Exploring Agriscience, 5th Edition.*

Introduction

Though it is still many years away for most of you, choosing a career path is something that you will all take part in. Choosing the career that is right for you depends on several factors. These include your personal and professional interests, your personality traits, your strengths (e.g., communication skills), your areas for development, and your overall goals. Most of you probably have not spent a great deal of time thinking about these things; however, exploring these options is one of the purposes of *Exploring Agriscience.*

Job and Career Opportunities

Having a vision or clear focus that points you in a particular direction can make that eventual choice easier for you and your family as it may help you eliminate things that may not fit within your future job selection criteria or particular interests. This activity allows you to do some preliminary job and career searching tasks and allows you to see what positions are available, beyond what is listed in this text. You conduct your own searches either via the Internet or (if the Internet is not readily accessible) via various magazines—old or new. The goal is not to find a job right now but to learn more about the field you are interested in, and what initial steps will be involved in a job search. A worksheet is included for recording your answers.

Activity: Job and Career Search

Materials Needed

- Computer, tablet, or smartphone with Internet access; search engine (e.g., Yahoo!, Google)
- Career worksheet (provided)

Procedure

1. Start with your device on and your browser started (e.g., Internet Explorer, Google Chrome), and go to a search engine site such as www.Google.com, www.Yahoo.com, or www.msn.com.
2. In the search box, type in a brief description of what you might be looking for. An example might be Agricultural Jobs, Careers in Agriculture, or Ag Careers. After typing your search criteria, hit the search button. The results will be posted based on popularity of links and how closely the content matches your search description.
3. Explore between 5 and 10 sites that are listed for you. Collect notes on three to five job descriptions that you find in these various sites. Each description should be based on a different job in various organizations or branches of the agriculture industry. For example, do not collect descriptions of the same position name within three different organizations; rather, select three different types of positions within those three organizations.
4. When you find those three which are of most interest to you, conduct another Internet search on those specific job descriptions, word for word (e.g., beef cattle farmer). Include in your notes any information regarding these positions' key responsibilities and qualifications. Make note of the things that you think you would enjoy most in each of those positions to share with your classmates later.

5. For your three positions of choice, find three sites or more that list directions or options for potential job seekers, such as positions available (regardless of location). Pick the three sites that are most useful based on user-friendliness, overall visual appeal, how up-to-date they are, and additional resources that are available to the job searcher.
6. Lastly, from the three positions you selected, choose one and do another search to find postsecondary institutions (i.e., two-year college, vocational college, four-year university) where you might be able to obtain a postsecondary degree or diploma in your field of choice. Select up to three based on criteria that are appealing to you.
7. Share your results with your classmates either in small groups or as part of a class discussion.

Questions for Thought

1. What did you learn?

2. What part of the activity did you enjoy most? Least?

Career Worksheet

Name:______________________________ Date:______________

Job search websites:

- ______________________________
- ______________________________
- ______________________________

Job requirements/qualifications:

- ______________________________
  ____
  ____
  ____
- ______________________________
  ____
  ____
  ____
- ______________________________
  ____
  ____
  ____

Postsecondary schools (school, city, state):

- ______________________________
- ______________________________
- ______________________________

Additional Notes